AF358679

Calixarene Complexes: Synthesis, Properties and Applications II

Calixarene Complexes: Synthesis, Properties and Applications II

Editor

Paula M. Marcos

Basel • Beijing • Wuhan • Barcelona • Belgrade • Novi Sad • Cluj • Manchester

Editor
Paula M. Marcos
Faculdade de Farmácia
Universidade de Lisboa
Lisboa
Portugal

Editorial Office
MDPI
St. Alban-Anlage 66
4052 Basel, Switzerland

This is a reprint of articles from the Special Issue published online in the open access journal *Molecules* (ISSN 1420-3049) (available at: www.mdpi.com/journal/molecules/special_issues/calixarene_ii).

For citation purposes, cite each article independently as indicated on the article page online and as indicated below:

Lastname, A.A.; Lastname, B.B. Article Title. *Journal Name* **Year**, *Volume Number*, Page Range.

ISBN 978-3-7258-0174-9 (Hbk)
ISBN 978-3-7258-0173-2 (PDF)
doi.org/10.3390/books978-3-7258-0173-2

Contents

About the Editor . . . . **vii**

Preface . . . . **ix**

Alexandre S. Miranda, Paula M. Marcos, José R. Ascenso, Mário N. Berberan-Santos and Filipe Menezes
Anion Binding by Fluorescent Ureido-Hexahomotrioxacalix[3]arene Receptors: An NMR, Absorption and Emission Spectroscopic Study
Reprinted from: *Molecules* **2022**, *27*, 3247, doi:10.3390/molecules27103247 . . . . **1**

Varun Rawat and Arkadi Vigalok
Electronic Tuning of Host-Guest Interactions within the Cavities of Fluorophore-Appended Calix[4]arenes
Reprinted from: *Molecules* **2022**, *27*, 5689, doi:10.3390/molecules27175689 . . . . **17**

Martin Tlustý, Václav Eigner, Hana Dvořáková and Pavel Lhoták
The Formation of Inherently Chiral Calix[4]quinolines by Doebner–Miller Reaction of Aldehydes and Aminocalixarenes
Reprinted from: *Molecules* **2022**, *27*, 8545, doi:10.3390/molecules27238545 . . . . **30**

Farida Galieva, Mohamed Khalifa, Zaliya Akhmetzyanova, Diana Mironova, Vladimir Burilov and Svetlana Solovieva et al.
New Supramolecular Hypoxia-Sensitive Complexes Based on Azo-Thiacalixarene
Reprinted from: *Molecules* **2023**, *28*, 466, doi:10.3390/molecules28020466 . . . . **44**

Loredana Ferreri, Marco Rapisarda, Melania Leanza, Cristina Munzone, Nicola D'Antona and Grazia Maria Letizia Consoli et al.
Calix[4]arene Derivative for Iodine Capture and Effect on Leaching of Iodine through Packaging
Reprinted from: *Molecules* **2023**, *28*, 1869, doi:10.3390/molecules28041869 . . . . **55**

Alexandre S. Miranda, Paula M. Marcos, José R. Ascenso, Mário N. Berberan-Santos, Peter J. Cragg and Rachel Schurhammer et al.
Critical Analysis of Association Constants between Calixarenes and Nitroaromatic Compounds Obtained by Fluorescence. Implications for Explosives Sensing
Reprinted from: *Molecules* **2023**, *28*, 3052, doi:10.3390/molecules28073052 . . . . **72**

Laura Baldini, Davide Balestri, Luciano Marchiò and Alessandro Casnati
A Combined Solution and Solid-State Study on the Tautomerism of an Azocalix[4]arene Chromoionophore
Reprinted from: *Molecules* **2023**, *28*, 4704, doi:10.3390/molecules28124704 . . . . **87**

Veronica Iuliano, Carmen Talotta, Paolo Della Sala, Margherita De Rosa, Annunziata Soriente and Placido Neri et al.
Hexahexyloxycalix[6]arene, a Conformationally Adaptive Host for the Complexation of Linear and Branched Alkylammonium Guests
Reprinted from: *Molecules* **2023**, *28*, 4749, doi:10.3390/molecules28124749 . . . . **100**

Shofiur Rahman, Mahmoud A. Al-Gawati, Fatimah S. Alfaifi, Wadha Khalaf Alenazi, Nahed Alarifi and Hamad Albrithen et al.
Detection of Aromatic Hydrocarbons in Aqueous Solutions Using Quartz Tuning Fork Sensors Modified with Calix[4]arene Methoxy Ester Self-Assembled Monolayers: Experimental and Density Functional Theory Study
Reprinted from: *Molecules* **2023**, *28*, 6808, doi:10.3390/molecules28196808 . . . . **113**

Zhongrui Chen, Gabriel Canard, Olivier Grauby, Benjamin Mourot and Olivier Siri
Breaking Azacalix[4]arenes into Induline Derivatives
Reprinted from: *Molecules* **2023**, *28*, 8113, doi:10.3390/molecules28248113 . . . . **128**

About the Editor

Paula M. Marcos

Paula M. Marcos graduated in Chemistry from the Faculdade de Ciências da Universidade de Lisboa (FCUL), and earned her PhD in Organic Chemistry in 1998 from the same University. She is a Professor at the Faculdade de Farmácia da Universidade de Lisboa (FFUL), and is a Researcher at the Centro de Química Estrutural (CQE). Currently, she is a member of the Topical Advisory Panel of Molecules for the Organic Chemistry section and a Review Editor on the Editorial Board of the Supramolecular Chemistry section of *Frontiers in Chemistry*. Her research is focused on macrocyclic and supramolecular chemistry, calixarene compounds, host–guest chemistry, ion and organic ion-pair recognition (including metal and organic cations, biologically and environmentally relevant anions, and biogenic amines and amino acids), NMR, and UV-Vis absorption spectroscopy.

Preface

Ion recognition continues to attract the interest of researchers all over the world due to the important role that cations and anions play in biological and chemical systems and in the environment. Calixarenes, owing to their structural features, are one of the most widely studied supramolecular hosts. These cyclic oligomers show important host–guest properties, which has led to numerous applications in a broad range of fields, including organocatalysis, sensing, extraction and separation, and more recently, biomedical applications. The relatively easy functionalization of their upper and lower rims, as well as the presence of a pre-organized cavity that is available in different sizes and conformations, make calixarenes attractive building blocks for the construction of supramolecular assemblies.

Due to the high participation of the scientific community in the previous Special Issue of Calixarene Complexes: Synthesis, Properties and Applications, a second one was successfully undertaken, and this new reprint presents the most recent developments in the calixarene field, including host–guest properties, as well as new synthetic methods and applications.

Paula M. Marcos
Editor

Article

Anion Binding by Fluorescent Ureido-Hexahomotrioxacalix[3]arene Receptors: An NMR, Absorption and Emission Spectroscopic Study

Alexandre S. Miranda [1,2], Paula M. Marcos [1,3,*], José R. Ascenso [4], Mário N. Berberan-Santos [2,*] and Filipe Menezes [5]

[1] Centro de Química Estrutural, Institute of Molecular Sciences, Faculdade de Ciências, Universidade de Lisboa, Edifício C8, 1749-016 Lisboa, Portugal; miranda.m.alexandre@gmail.com
[2] IBB-Institute for Bioengineering and Biosciences, Instituto Superior Técnico, Universidade de Lisboa, 1049-001 Lisboa, Portugal
[3] Faculdade de Farmácia, Universidade de Lisboa, Av. Prof. Gama Pinto, 1649-003 Lisboa, Portugal
[4] Centro de Química Estrutural, Institute of Molecular Sciences, Instituto Superior Técnico, Complexo I, Av. Rovisco Pais, 1049-001 Lisboa, Portugal; jose.ascenso@ist.utl.pt
[5] Institute of Structural Biology, Helmholtz Zentrum Muenchen, Ingolstaedter Landstr. 1, 85764 Neuherberg, Germany; filipemmenezes@gmail.com
* Correspondence: pmmarcos@fc.ul.pt (P.M.M.); berberan@tecnico.ulisboa.pt (M.N.B.-S.)

Abstract: Fluorescent receptors (**4a**–**4c**) based on (thio)ureido-functionalized hexahomotrioxacalix[3]arenes were synthesised and obtained in the partial cone conformation in solution. Naphthyl or pyrenyl fluorogenic units were introduced at the lower rim of the calixarene skeleton via a butyl spacer. The binding of biologically and environmentally relevant anions was studied with NMR, UV–vis absorption, and fluorescence titrations. Fluorescence of the pyrenyl receptor **4c** displays both monomer and excimer fluorescence. The thermodynamics of complexation was determined in acetonitrile and was entropy-driven. Computational studies were also performed to bring further insight into the binding process. The data showed that association constants increase with the anion basicity, and AcO^-, BzO^- and F^- were the best bound anions for all receptors. Pyrenylurea **4c** is a slightly better receptor than naphthylurea **4a**, and both are more efficient than naphthyl thiourea **4b**. In addition, ureas **4a** and **4c** were also tested as ditopic receptors in the recognition of alkylammonium salts.

Keywords: hexahomotrioxacalix[3]arenes; fluorescent anion receptors; NMR studies; UV–Vis absorption studies; fluorescence studies; theoretical calculations

Citation: Miranda, A.S.; Marcos, P.M.; Ascenso, J.R.; Berberan-Santos, M.N.; Menezes, F. Anion Binding by Fluorescent Ureido-Hexahomotrioxacalix[3]arene Receptors: An NMR, Absorption and Emission Spectroscopic Study. *Molecules* **2022**, *27*, 3247. https://doi.org/10.3390/molecules27103247

Academic Editor: Ana Margarida Gomes da Silva

Received: 1 May 2022
Accepted: 16 May 2022
Published: 19 May 2022

Publisher's Note: MDPI stays neutral with regard to jurisdictional claims in published maps and institutional affiliations.

1. Introduction

Fluorescence spectroscopy, due to its high sensitivity and simplicity, is an attractive technique used for the quantitative determination of ions. Lately, a wide range of fluorescence sensors based on calixarenes have been used in various applications, namely in the detection of biologically and environmentally relevant cations and anions [1–3]. Owing to their structural features, these macrocycle compounds have been widely investigated as neutral molecule and ion receptors [4,5]. They possess a well-defined hydrophobic cavity available in different sizes and conformations, and an almost unlimited number of derivatives can be obtained by functionalisation of their upper and lower rims. Different fluorophores like naphthalene, anthracene and pyrene are among the most incorporated in the calixarene framework, leading to fluorescent probes for the recognition of different types of analytes. Examples of fluorescent calix[4]arene [6–13], calix[5]arene [14] and calix[6]arene [15,16] receptors have been reported in the literature.

Anions [17–19] such as fluoride, chloride, and iodide play important roles in many medical systems, and the carboxylate group is present in several biological molecules.

Concerning the environment, the treatment of nitrates, sulphates, phosphates, and perchlorates, among other anions, is also a demanding field. Synthetic anion receptors, namely calixarenes bearing (thio)urea moieties, can form strong and directional hydrogen bonds between their NH groups and the anions. Some of these calixarenes also behave as ditopic receptors [20], simultaneously binding both ions of a given ion pair.

As part of our continuous interest on the host–guest properties of homooxacalixarenes (calixarene analogues in which the CH_2 bridges are partly or completely replaced by CH_2OCH_2 groups) [21,22], we have reported the anion and ion pair recognition of several (thio)ureido-dihomooxacalix[4]arenes [23–25], including fluorescent receptors [26,27]. The ion affinity of hexahomotrioxacalix[3]arenes, a very interesting macrocycle formed by an 18-membered ring and having only two basic conformations, has also been investigated [28–33]. Recently, we have extended our research into the study of fluorescent hexahomotrioxacalix[3]arenes. Thus, this paper describes the synthesis of three new fluorescent derivatives bearing naphthylurea (**4a**), naphthylthiourea (**4b**), and pyrenylurea (**4c**) residues at the lower rim via a butyl spacer. These compounds were obtained in a partial cone conformation in solution. Their affinity towards relevant anions and alkylammonium salts was assessed by proton NMR, UV–Vis absorption and fluorescence spectroscopy. The thermodynamics of the anion complexation was determined in acetonitrile by absorption and emission, and computational studies were also performed to bring further insight about the binding process.

2. Results and Discussion

2.1. Synthesis and Conformational Analysis

Recently, we have reported the synthesis of three hexahomotrioxacalix[3]arene- based receptors bearing (thio)urea groups linked to the macrocycle lower rim by a butyl spacer [32]. Following this line of research, we prepared three new fluorescent sensors containing (thio)urea residues and naphthalene or pyrene moieties. Thus, a three-step procedure (already described) [32] was undertaken from parent compound *p-tert*-butylhexahomotrioxacalix[3]arene **1**. The alkylation reaction of **1** with *N*-(4-bromobutyl)phthalimide and K_2CO_3 afforded triphthalimide compound **2**. The phthalimido groups were further removed with hydrazine, yielding triamine **3**, which reacted with naphthyliso(thio)cyanate or 1-pyrenylisocyanate to give the corresponding tri(thio)urea receptors (naphthyl urea **4a**, naphthyl thiourea **4b** and pyrenyl urea **4c**) in the partial cone conformation (Scheme 1).

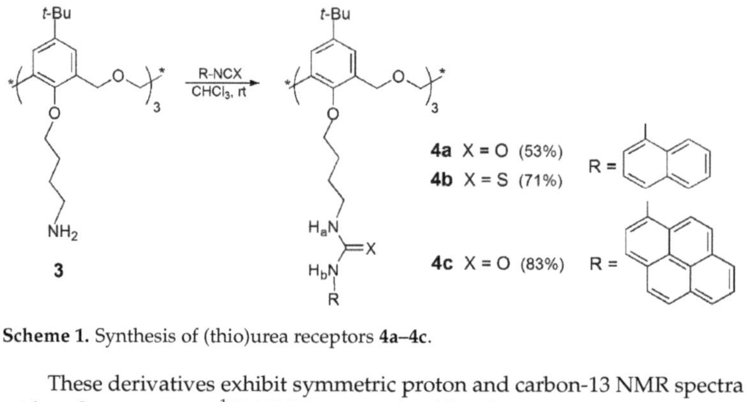

Scheme 1. Synthesis of (thio)urea receptors **4a–4c**.

These derivatives exhibit symmetric proton and carbon-13 NMR spectra compatible with a C_s symmetry. ^1H NMR spectra in chloroform at room temperature show two singlets (in a 1:2 ratio) for *tert*-butyl groups, three AB quartets for the CH_2 bridge protons (CH_2OCH_2), one pair of doublets and one singlet for the aromatic protons of the calixarene skeleton, two triplets and two singlets (in a 1:2 ratio) for the NH_a and NH_b protons, respectively, besides several multiplets for the $-OCH_2CH_2CH_2CH_2N-$ methylene protons and for the aromatic protons of the naphthyl/pyrenyl groups. The proton assignments were confirmed by COSY spectra. ^{13}C NMR spectra show 9 of the expected 10 upfield resonances arising from the *tert*-butyl groups and the methylene carbon atoms of the $-OCH_2CH_2CH_2CH_2N-$ group, 5 midfield resonances arising from the CH_2OCH_2 and the $-OCH_2CH_2CH_2CH_2N-$ groups, and 31 (or 41 in the case of **4c**) downfield resonances arising from the aromatic carbon atoms and the (thio)carbonyl groups. All resonances were assigned by DEPT experiments.

2.2. Anion and Ion-Pair Recognition

2.2.1. Proton NMR Studies

The binding properties of (thio)ureas **4a–4c** toward several relevant anions of spherical (F^-, Cl^-, Br^-, I^-), trigonal planar (NO_3^-, AcO^-, BzO^-) and tetrahedral (HSO_4^-, ClO_4^-) geometries were investigated using tetrabutylammonium (TBA) salts in $CDCl_3$ by proton NMR titrations. The association constants reported as log K_{ass} in Table 1 were determined following the NH chemical shifts of the (thio)ureido groups through the WinEQNMR2 programme [34].

Table 1. Association constants (log K_{ass}) [a] of hexahomotrioxa (thio)ureas **4a–4c** determined by ^1H NMR in $CDCl_3$ at 25 °C.

	Spherical				Trigonal Planar			Tetrahedral	
	F^-	Cl^-	Br^-	I^-	NO_3^-	AcO^-	BzO^-	HSO_4^-	ClO_4^-
I. Radius/Å [b]	1.33	1.81	1.96	2.20	1.79	2.32	—	1.90	2.00
Napht urea **4a**	2.91	2.88	2.33	1.83	2.29	3.20	3.09	2.70	1.57
Napht thiourea **4b**	2.73	2.00	1.23	1.11	1.40	2.79	2.61	2.05	1.18
Pyr urea **4c**	3.31	3.19	2.78	2.23	2.67	3.42	3.45	2.97	1.99

[a] Estimated error < 10%. [b] Data quoted in I. Marcus, *Ion Properties*, Marcel Dekker, New York, pp. 50–51, 1997.

The addition of TBA salts to the receptors resulted in downfield shifts of their NH protons, clearly indicating hydrogen bonding interactions between the (thio)urea groups and the anions (see, for example, Figure 1 for **4a** + BzO^-). A fast exchange rate between free and complexed receptors was observed on the NMR time scale at room temperature. The titration curves obtained (Figure S1) indicate the formation of 1:1 complexes, this stoichiometry being also confirmed by Job plots (Figure S2).

Data presented in Table 1 show that AcO^-, BzO^- and F^- are the best bound anions, and, in general, the association constants increase with increasing of anion basicity for all the receptors. A closer analysis of the results indicates that pyrenyl urea **4c** is a more efficient receptor than naphthyl urea **4a**, presenting association constants that are in average 0.35 log units higher than those obtained for **4a**. This fact may be due to the presence of the bulkier pyrenyl residue that makes receptor **4c** less flexible and consequently more preorganised compared to naphthyl urea **4a**. Concerning spherical halides, ureas **4a** and **4c** both form strong complexes with F^- (log K_{ass} = 2.91 and 3.31, respectively), and the association constants decrease with decreasing of anion basicity. Complexes with trigonal planar oxoanions AcO^- and BzO^- display the highest log K_{ass} values for **4a** (3.20 and 3.09, respectively) and for **4c** (3.42 and 3.45, respectively), and in the case of the tetrahedral inorganic oxoanion HSO_4^-, reasonably good association constants were also obtained.

Naphthyl thiourea **4b** was also studied, and the data reported in Table 1 show that **4b** is a weaker receptor than corresponding urea **4a** despite the increased acidity of its NH groups. Thiourea **4b** displays the same trend as urea **4a**, the anions being bound according to their basicity for all the geometry groups. The association constants were 0.85 log units lower on average than those obtained for **4a**, except for the more basic and best bound anions F^-, AcO^- and BzO^- (Δ log K_{ass} = 0.18, 0.41 and 0.48, respectively). Identical situations were reported for several homooxacalixarene thiourea receptors [23,27,32]. The larger size of the sulphur atom that distorts the *cis–cis* geometry required for anion binding may be the cause of this effect. As a result, less preorganised and energetically less favourable thiourea groups are obtained compared to the urea ones [35].

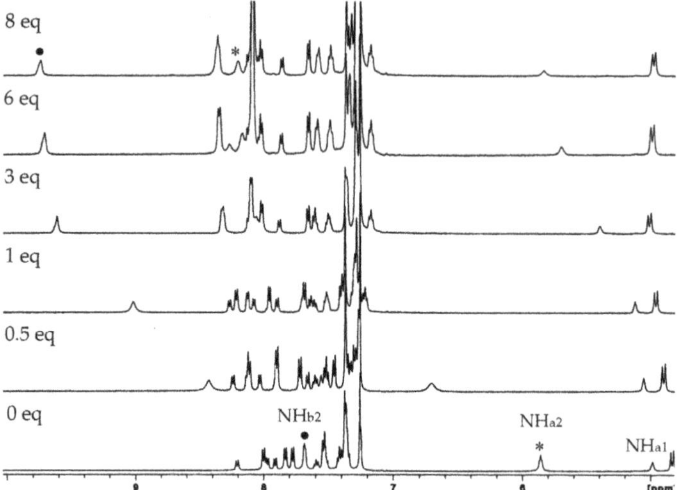

Figure 1. ^{1}H NMR partial spectra (500 MHz, CDCl$_3$, 25 °C) of Naphturea **4a** with several equiv of TBA BzO.

The calixarene skeletons of these partial cone ureas seem to undergo only slight conformational changes upon complexation, as observed before for other hexahomotrioxa derivative analogues [32]. Very small chemical shift variations were observed for the *tert*-butyl protons upon the addition of 8 equiv of the salts for both ureas ($\Delta\delta \leq 0.03$ ppm). One of the three aromatic protons also showed small downfield variations ($\Delta\delta \leq 0.07$ and 0.09 ppm for **4a** and **4c**, respectively), while the others are difficult to follow due to overlapping by the naphthyl/pyrenyl peaks.

Naphthyl and pyrenyl ureas **4a** and **4c** were also tested in the recognition of *n*-propyl and *n*-butylammonium chlorides in a preliminary study to evaluate their ability as ditopic receptors.

Proton NMR titrations were carried out by adding increasing amounts (up to two equiv) of the salts to CDCl$_3$ solutions of receptors **4a** and **4c**. In the case of **4a**, the addition of one equiv of both guests at room temperature caused broadening of the peaks; while for **4c** this broadening was also followed by the appearance of resonances at high field. This indicates host–guest interaction with both receptors, and the inclusion of alkyl groups inside the aromatic cavity of host **4c** at room temperature. To better analyse the spectra, the temperature was lowered to 233 K. For both ureas, peaks have appeared in the negative region of the spectra (Figure S3), but they remain broad, suggesting that the complexation/decomplexation process is still fast on the NMR time scale. Given these results, no other ion pairs were tested.

2.2.2. UV–Vis Absorption and Fluorescence Studies

The proton NMR studies of the interactions between (thio)ureas **4a**–**4c** and the previous anions were complemented by UV–Vis absorption and fluorescence titrations in dichloromethane, and also in acetonitrile in the case of urea **4a**.

Naphthylurea **4a** behaved similarly in both solvents, displaying an absorption band centred at approximately 301 or 295 nm in CH$_2$Cl$_2$ or MeCN, respectively, but with a marked shoulder at 330 nm and an onset at about 360 nm. The low-energy part of the absorption (wavelengths larger than about 300 nm) mainly arises from the naphthalene moieties, as shown by a comparison with both the spectrum of parent calixarene 1 whose lowest energy absorption is benzenoid, peaking at 282 nm and with an onset at about 300 nm, and with the absorption spectrum of naphthyl precursor **A**, whose absorption peaks at 300 nm, displays a shoulder at 330 nm, and had the onset at about 360 nm (Figure 2a). Upon the addition of increasing amounts of F^-, Cl^-, AcO^- and BzO^-, this band decreases in intensity while a new one progressively appears at longer wavelengths. In the case of F^-, the new maximum is reached at approximately 315 or 308 nm, with red shifts of 14 or 13 nm, respectively in CH_2Cl_2 or MeCN. Isosbestic points at 263 and 299 nm are observed in both solvents (Figure 3). For the other anions, similar absorption spectral changes were observed (Figure S4), showing also isosbestic points and red shifts, albeit smaller (from 9 to 6 nm). Concerning Br^-, NO_3^-, HSO_4^- and ClO_4^- anions, their complexation by receptor **4a** induced successive increases of the absorption (except for Br^- in MeCN that decreases as the anion concentration increase), but with no wavelength shift of the absorption maximum (Figure S5). Naphthyl thiourea **4b** showed a different behaviour for all the anions. In the case of F^-, two isosbestic points at approximately 277 and 309 nm can be seen, but no significant shift of the absorption maximum is observed (Figure S6a). For the other anions, the absorption band decreases as the anion concentration increases, presenting an isosbestic point around 325 nm (Figure S6b).

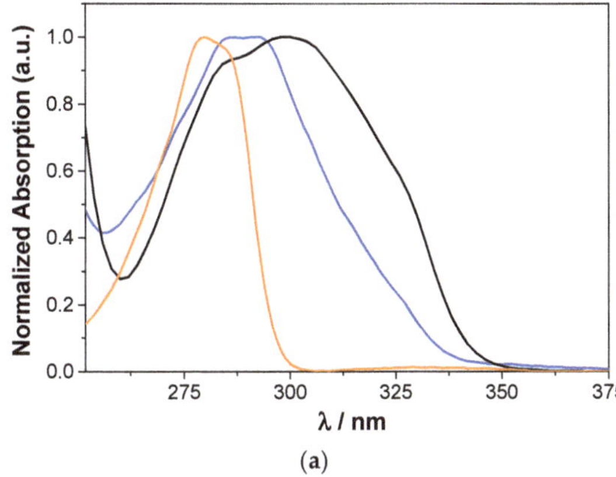

Free pyrenyl urea **4c** displays two absorption bands of similar intensity peaking at approximately 283 and 346 nm, both arising from the pyrenyl groups, as shown by comparison with the absorption of the isolated moiety precursor **B** (Figure 2b) and again considering the benzenoid absorption characteristics of the calixarene backbone. The titration of **4c** with F^-, Cl^-, AcO^- and BzO^- anions resulted in an increase of the intensity of both bands with slight bathochromic shifts of their absorption maxima (Figure 4). These shifts were higher for the long wavelength pyrenyl band ($\Delta\lambda = 12$ nm for F^- and 4 nm for Cl^-) compared to the macrocycle one ($\Delta\lambda = 3$ nm for F^- and 1 nm for Cl^-). Additions of Br^-, NO_3^-, HSO_4^- and ClO_4^- anions to **4c** also induced progressive increases in the absorption bands, but no shifts in their maxima were observed (Figure S7).

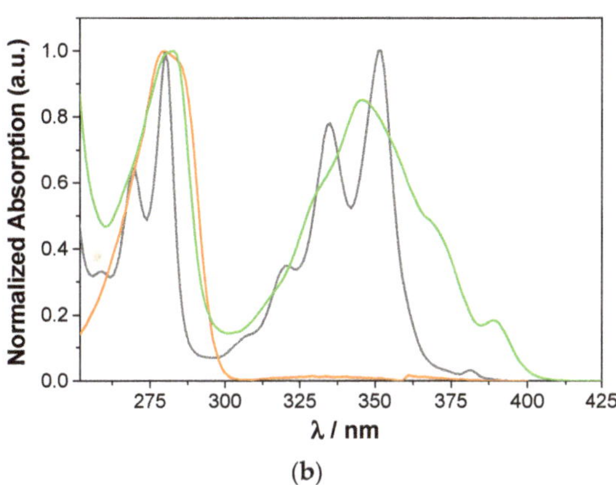

Figure 2. Normalised absorption spectra of (a) **1** (orange), **A** (blue) and **4a** (black) and (b) **1** (orange), **B** (grey) and **4c** (green) in CH_2Cl_2 at 25 °C. [1] = [A] = 5×10^{-5} M, [4a] = 2×10^{-5} M; [B] = 4×10^{-5} M, [4c] = 1×10^{-5} M.

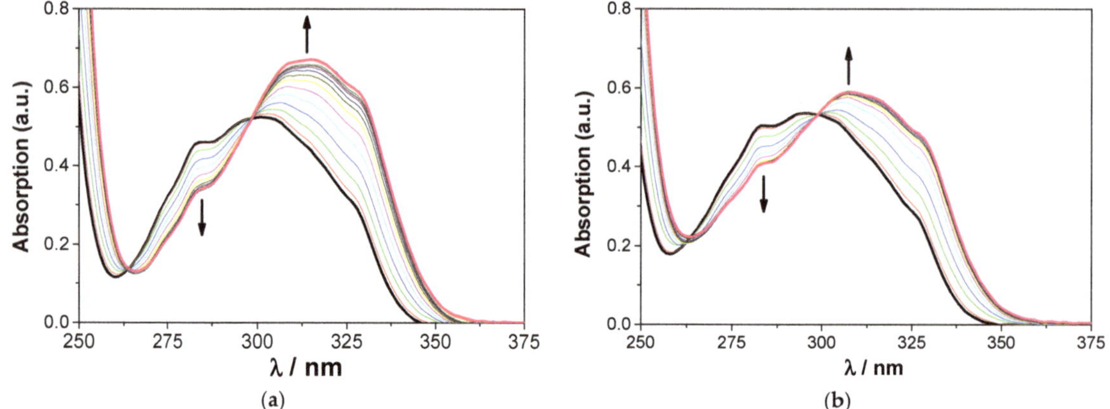

Figure 3. Changes in the UV spectrum of urea **4a** (2.0×10^{-5} M) upon addition of TBA F (up to 10 equiv.) in (**a**) CH_2Cl_2 and (**b**) MeCN. The arrows indicate the effect of increasing amounts of salt.

Naphthylurea **4a** presents emission bands centred at approximately 375 nm in both solvents, as well as nanosecond fluorescence lifetimes and significant quantum yields (Table 2 and Figure S8). No intramolecular excimer is observed in the fluorescence spectra, pointing to the impossibility of naphthyl group pairs to attain near-sandwich configurations. An increase of the emission intensity was obtained upon the addition of all anions to **4a** in CH_2Cl_2 (Figure 5a), which was less pronounced for NO_3^-, HSO_4^- and ClO_4^- (Figure S9a). In MeCN, an increase of the emission intensity was also observed in the case of Cl^-, Br^-, AcO^-, BzO^- and ClO_4^- anions, whereas a quenching of the fluorescence intensity was observed for F^- (Figure 5b), NO_3^- and HSO_4^- (Figure S9b). The lifetimes of **4a** either do not change or change moderately upon complexation (Table S1). At 10 equivalents, AcO^-, BzO^- and F^- have the strongest effect on the lifetimes of **4a**, in agreement with the expected associated fractions (see Figure S10). The observed changes both in lifetimes and in intensities (hence fluorescence quantum yields) probably resulted from conformational modifications mainly affecting the radiative constants.

Table 2. Photophysical properties of model compounds **A** and **B** and ureas **4a** and **4c** at 25 °C.

	Solvent	$\lambda_{max,abs}$ (nm)	ε (M^{-1} cm^{-1})	$\lambda_{max,em}$ (nm)	Stokes Shift [a] (nm)	τ_f (ns)	Φ_F [b]	k_r (ns^{-1})	k_{nr} (ns^{-1})
A	CH_2Cl_2	293	2.5×10^4	372	79	2.61	0.52	0.20	0.18
B	CH_2Cl_2	352	4.7×10^4	412	60	5.70	0.14	0.025	0.15
4a	CH_2Cl_2	301	5.1×10^4	375	74	4.84	0.37	0.076	0.13
	MeCN	295	5.6×10^4	374	79	3.67	0.35	0.095	0.18
4c	CH_2Cl_2	283 [c]	5.6×10^4	398 [e]	115	1.98 [g]	0.19 [i]	—	—
		340 [d]		498 [f]	158	25.9 [h]		—	—

[a] Computed as $\lambda_{max,em} - \lambda_{max,abs}$; [b] Against quinine sulfate $\Phi_F = 0.546$ in H_2SO_4 0.5 M; [c] Abs_{max} (calixarenic macrocycle + pyrene band); [d] Abs_{max} (pyrene band); [e] Corresponding to the monomer band; [f] corresponding to the excimer band; [g] $\lambda_{em} = 398$ nm; [h] $\lambda_{em} = 498$ nm; [i] overall (monomer + excimer) emission.

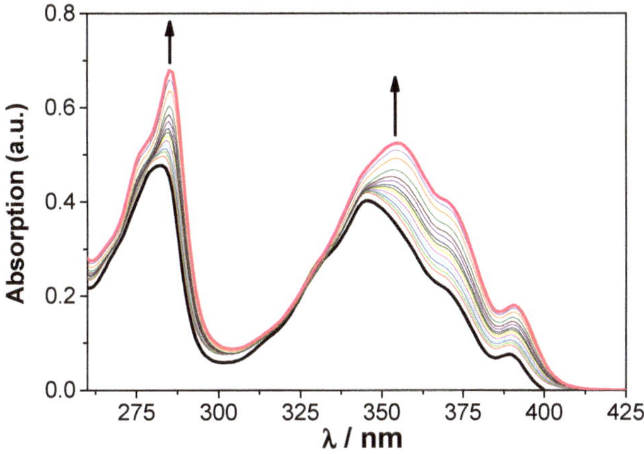

Figure 4. Changes in UV spectrum of urea **4c** (1.0×10^{-5} M) upon the addition of TBA AcO (up to 10 equiv.) in CH_2Cl_2. Arrows indicate increasing amounts of salt.

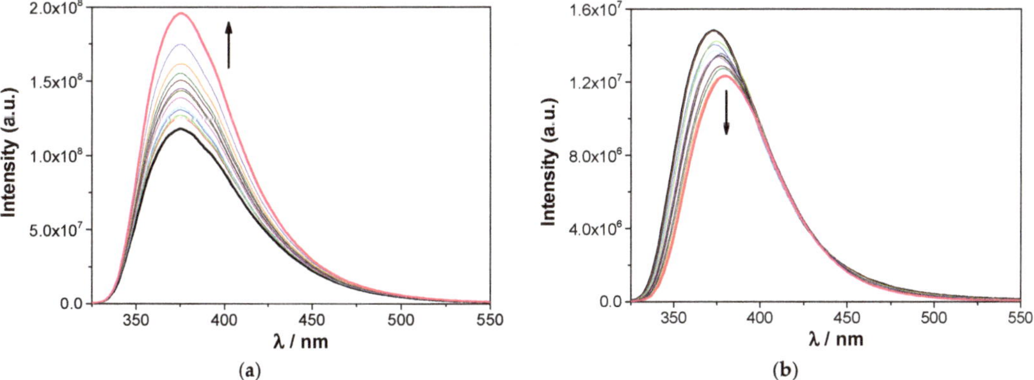

(a)

(b)

Figure 5. Changes in the emission spectrum of urea **4a** (2.0×10^{-5} M) upon addition of TBA F (up to 10 equiv) in (**a**) CH_2Cl_2 and (**b**) MeCN. The arrows indicate the effect of increasing amounts of salt.

The fluorescence of thiourea **4b** could not be studied, as this derivative is unstable upon irradiation.

The fluorescence of **4c** exhibits both monomer (ca. 400 nm) and excimer (ca. 500 nm) bands arising from the pyrene moieties (Table 2 and Figure S8). The overall fluorescence quantum yield is significant, and monomer and excimer have very different lifetimes, the average monomer lifetime being of the order of 2 ns, significantly shorter than that of reference compound **B** (Table 2), whereas the excimer lifetime is one order of magnitude higher. Owing to the close proximity of pyrene pairs, the excimer displays a very short risetime (Table S2), indicating that its formation is quite fast. This component is also found in the monomer decay (Table S2) along with a component close to that of reference compound **B**. This probably results from the fact that two pyrenyl units engage in excimer formation, whereas the third, on account of the partial cone conformation, will exhibit isolated monomer behavior. The monomer–excimer double emission, with its associated excited-state kinetics and photophysics, precludes a clear-cut evaluation of association constants from fluorescence. The average monomer lifetimes of **4c** do not change significantly upon complexation (Table S2). It is nevertheless observed that complexation has a varying effect on excimer photophysics depending on the anion (Figure 6a), showing that different conformations of the pyrenylurea arms can be adopted upon complexation. It is observed that in some cases the excimer/monomer intensity ratio is essentially unaffected by anion binding (Figure S11), whereas in other cases there is a strong effect (Figure 6b).

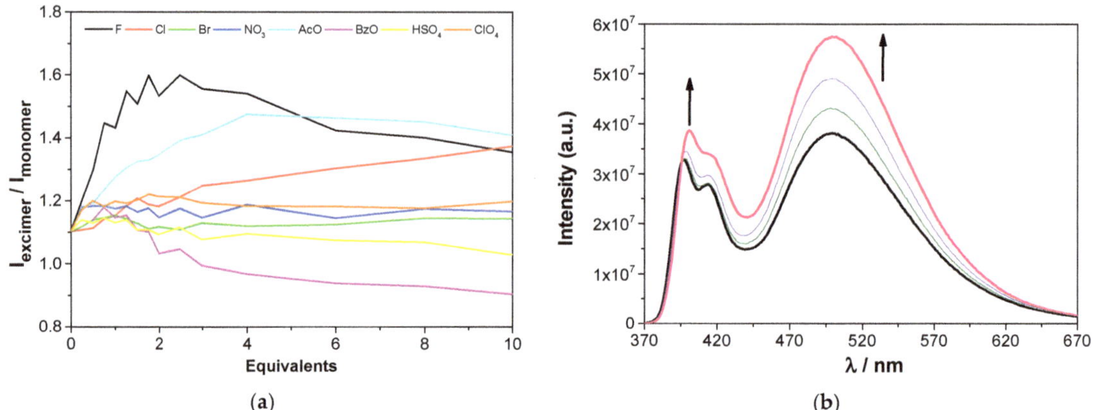

(a)   (b)

Figure 6. (a) Ratio of excimer and monomer peak intensities of urea **4c** upon addition of TBA salts, (b) Changes in the emission spectrum of urea **4c** upon addition of 0 (black), 0.75 (olive), 2 (violet) and 10 equiv. (pink) of TBA AcO. The arrows indicate the increasing amounts of salt. [**4c**] = 1.0×10^{-5} M in CH_2Cl_2.

In all cases, significant spectral variations were observed for the three receptors, allowing the determination of the corresponding binding constants by absorption in dichloromethane (Table 3). For naphthyl urea **4a**, the constants were also determined in dichloromethane and acetonitrile, and using both absorption and fluorescence. The association constants obtained for all receptors are on average one log unit higher than those obtained by NMR, but follow approximately the same trend. The more dilute solutions used in the UV–Vis/fluorescence titrations favour the dissociation of the salts, providing a higher concentration of the anions available for complexation and resulting in higher association constants [14]. The data reported in Table 3 show a stronger binding in dichloromethane than that in acetonitrile for urea **4a**, both by absorption and emission, in agreement with the relative competitive character of the two solvents. As observed before (NMR studies), pyrenyl urea **4c** is a slightly better receptor than naphthyl urea **4a** for all the anions, while naphthyl thiourea **4b** is the weakest receptor.

Table 3. Association constants (log K_{ass}) [a] of (thio)ureas **4a–4c** determined by UV–Vis absorption and emission at 25 °C.

			Spherical			Trigonal Planar			Tetrahedral	
		Solvent	F^-	Cl^-	Br^-	NO_3^-	AcO^-	BzO^-	HSO_4^-	ClO_4^-
4a	Abs	CH_2Cl_2	4.34	3.74	3.43	3.51	4.23	4.28	3.65	2.69
		MeCN	4.25	3.51	3.26	3.28	4.04	4.01	3.54	2.54
	Emi	CH_2Cl_2	4.20	3.78	3.29	3.44	4.07	4.13	3.45	2.65
		MeCN	4.14	3.46	3.18	3.19	3.91	4.06	3.29	2.48
4b	Abs	CH_2Cl_2	3.39	3.00	2.78	2.62	3.28	3.19	2.91	2.26
4c	Abs	CH_2Cl_2	4.46	3.97	3.65	3.60	4.29	4.37	3.80	2.92

[a] Estimated error < 10%.

2.2.3. Temperature Dependence of the Association Constants—Thermodynamic Analysis and Simulation

The thermodynamic parameters of the complexation of naphthylurea **4a** with F^- and AcO^- in acetonitrile were determined from the dependence of the association constant with temperature using the van't Hoff equation (Figure 7 and Table 4). According to Table 4, similar results were obtained from both experimental techniques. A general conclusion drawn from the data is that the complexation is entropy-driven in the two cases studied, which is surprising and was only observed once in similar systems [36].

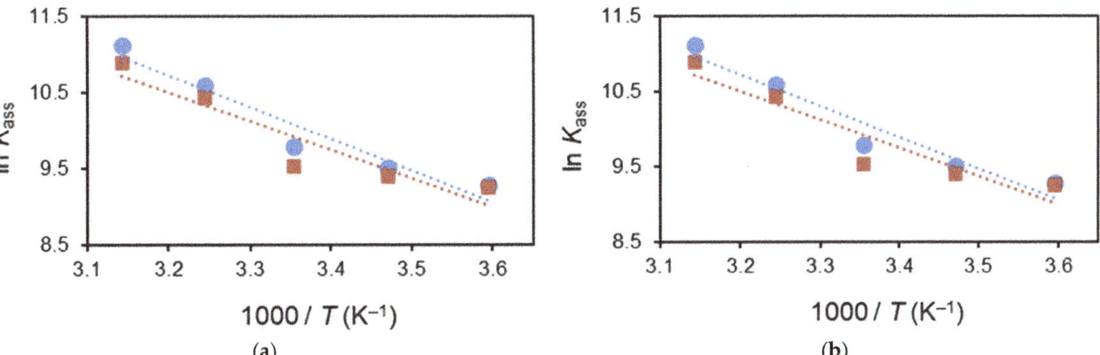

Figure 7. Linear regression of ln K_{ass} vs. $1000/T$ using UV–Vis absorption (circles) and fluorescence (squares) for urea **4a** in MeCN (2.0×10^{-5} M) upon addition up to 10 equiv of (**a**) TBA F and (**b**) TBA AcO.

Table 4. Thermodynamic quantities for the complexation of anions with **4a** in MeCN at 25 °C determined by UV–Vis and fluorescence.

	log K_{ass}	$\Delta G°$ (kJ mol^{-1})	$\Delta H°$ (kJ mol^{-1})	$\Delta S°$ (J mol^{-1}K^{-1})
F$^-$	4.25/4.14	$-25/-25$	35/32	200/190
AcO$^-$	4.04/3.91	$-23/-22$	35/32	190/180

To better understand the entropy effect, extensive simulations were carried out. For this purpose, several main conformations of naphthylurea **4a** arms (oriented on the same side of the macrocyclic ring) were identified, according to the orientation of each urea NH and naphthyl moieties. Each urea group may exist in a *sin* (S, both protons in the same direction) or in an *anti* (A) orientation, and the naphthyl groups may be π-stacked (structures marked with a prime or identified by dashed lines, as shown in Figure 8a) or free (unprimed structures). Figure 8a shows the relative chemical potentials of these species in acetonitrile solutions, and in Figure 8b is represented the optimised structure for the SS' conformer. Gas phase thermodynamic data is provided in the Supporting Information.

As expected, π-stacked structures are not as stable in solution as the respective open forms, mainly due to the reduced solvent accessible surface areas. The exception is conformer SS', which benefits from intramolecular hydrogen bonding as well (see Figure 8b). The open form of conformer SA is also reasonably close in chemical potential to conformers SS and SS'. SA' on the other hand, is quite unstable in solution. Unless there are kinetic factors hindering the interconversion between different species, equilibrium solutions of the calixarene are expected to be dominated by SS and SS', with a combined mole fraction of 95%. The two SA conformers have a weight of 4%, whereas the AA corresponds to only 1%. The combined estimation of weights is based on the conformational search performed with the program CREST, which showed that primed and unprimed species belong to the same ensemble.

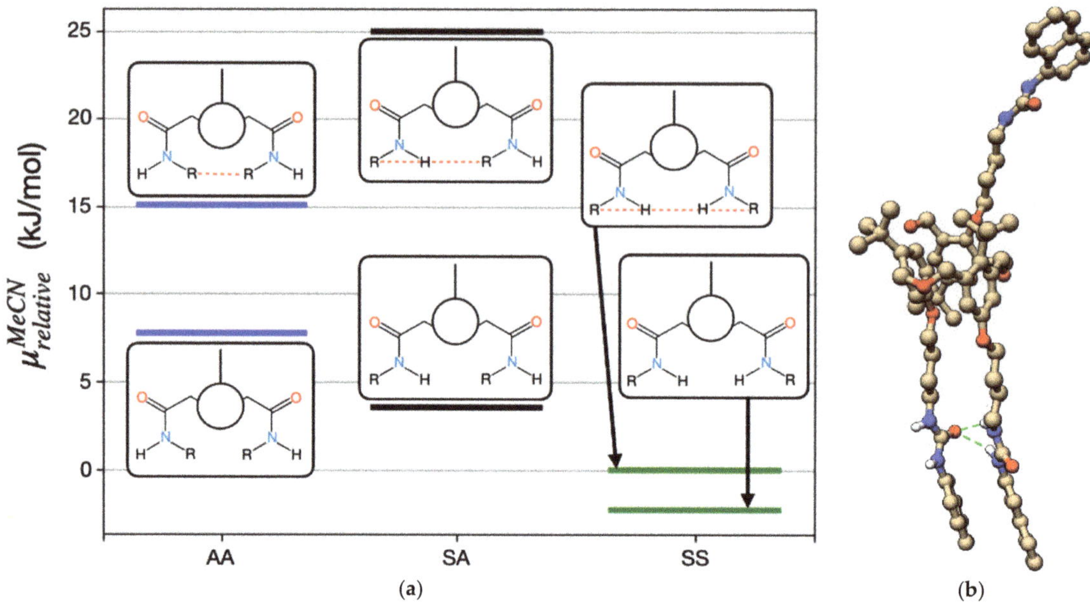

Figure 8. (**a**) Relative chemical potentials of **4a** arm main conformers in MeCN (as a continuum dielectric) at 300 K. Dashed lines denote π-stacking interactions of the naphthyl groups; (**b**) optimised structure for the SS′ conformer. Intramolecular hydrogen bonds are represented by green dashed lines.

Calculated Gibbs energy variations upon binding of F^- anion in acetonitrile for conformers SS are displayed in Figure 9a. This figure shows results where the solvent participation is implicit ($n = 0$), as well as hybrid models, where 1 to 4 molecules of acetonitrile are explicitly included in the system (besides the implicit effects from the generalised Born model). These molecules are initially attached to the urea moieties by hydrogen bonding. Calculated gas-phase enthalpies (Figure S12a) and entropies (Figure 9b) significantly differed from experimental data, stressing the role of the solvent. The gas-phase calculations predict enthalpies from -123 up to -83 kJ mol^{-1} and entropy variations also negative from -206 to -230 J mol^{-1} K^{-1} for conformers SS, as one could expect for a binding process. The experimental average values are positive and of the order of 33 kJ mol^{-1} and 195 J mol^{-1} K^{-1}, respectively (Table 4). The results for the entropy variations considering an estimation of the conformational effects for conformer SS are also shown in Figure 9b (denoted as SSconf). These results evidence the relevance of considering the loss of conformational degrees of freedom when the calixarene binds the anion.

The designed hybrid models help to clarify the role of acetonitrile in the binding process. Instead of binding, the formation of adducts between fluoride and **4a** is transformed into exchange reactions. Acetonitrile may bind the calixarene by hydrogen bonding or dispersion. Though the enthalpy gain for the latter type of interaction is weaker, entropic penalties for both binding modes are similar. Irrespective of the conformer considered in the binding process, enthalpies and entropies progressively increase with the number of added acetonitrile molecules. On the other hand, Gibbs energies progressively decrease when acetonitrile molecules are explicitly included.

The obtained results clearly show that the explicit inclusion of acetonitrile solvent (as a few discrete molecules) brings the computed enthalpy and entropy closer to the measured values. Furthermore, inclusion of conformational effects is essential for the correct prediction of the entropy changes, keeping this state function from reaching unrealistically large values. The positive enthalpies and entropies observed experimentally result therefore from the direct involvement of the solvent in the binding process.

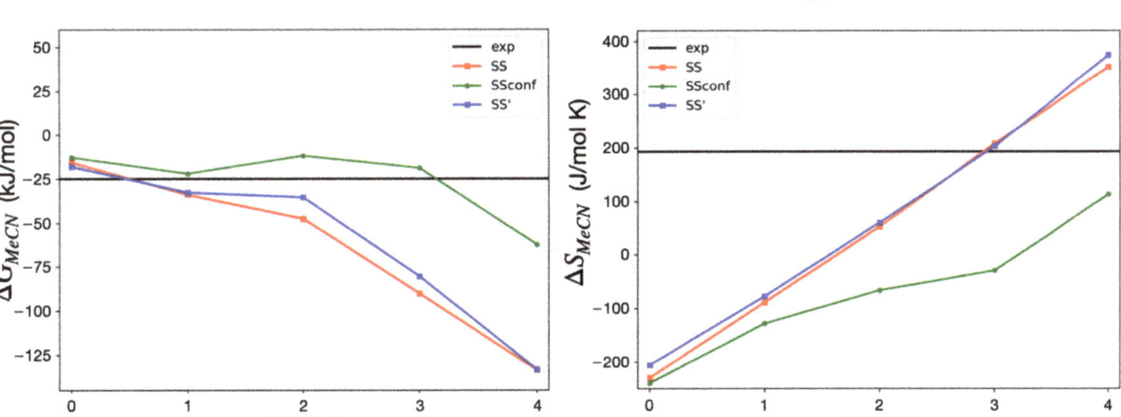

Figure 9. Variation in (**a**) Gibbs energy and (**b**) entropy for the binding F^- to **4a** with n molecules ($n = 0$–4) of MeCN explicitly considered. The reactions considered to generate the plot are summarised by, e.g., $SS : nMeCN + F^- \rightleftharpoons nMeCN + SS : F^-$.

3. Experiments

3.1. General Information

All chemicals were reagent-grade and used without further purification. Chromatographic separations were performed on Merck silica gel 60 (particle size 40–63 µm, 230–400 mesh). Melting points were measured (not corrected) on a Stuart Scientific apparatus, and FTIR spectra were recorded on a Shimadzu Model IRaffinity-1 spectrophotometer. ^{1}H and ^{13}C NMR spectra were recorded on a Bruker Avance III 500 MHz spectrometer with TMS as internal reference. Conventional COSY 45 experiments were conducted as 256×2 K complex points. Elemental analysis was determined on a Fisons EA 1108 microanalyser.

3.2. Procedure for the Synthesis of Ureas **4a** and **4c**, and Thiourea **4b**

To a solution of amine **3** [32] (0.53 g, 0.67 mmol) in 30 mL of CHCl$_3$ was added 2.01 mmol of the appropriate iso(thio)cyanate. In the case of 1-pyrenylisocyanate, it was synthesised from 1-pyrene amine in the presence of triethylamine and triphosgene, according to reference [37]. The mixture was stirred at room temperature under N$_2$ for 4 h. Evaporation of the solvent yielded the crude products that were purified as described below.

7,15,23-Tri-tert-butyl-25,26,27-tri[[(N′-1-naphthylureido)butyl]oxy]-2,3,10,11,18,19-hexahomo-3,11,19-trioxacalix[3]arene (**4a**): Flash chromatography (SiO$_2$, eluent CH$_2$Cl$_2$/MeOH, from 99.5:0.5 to 99:1); it was obtained in 53% yield (0.46 g); m.p. 157–159 °C; IR (KBr) 3343 cm^{-1} (NH), 1647 cm^{-1} (CO); ^{1}H NMR (CDCl$_3$, 500 MHz) δ 0.68 (m, 4H, OCH$_2$C**H**$_2$CH$_2$CH$_2$-NH$_a$ inverted), 1.25 [s, 18H, C(CH$_3$)$_3$], 1.30 [s, 9H, C(CH$_3$)$_3$ inverted], 1.39 (m, 4H, OCH$_2$-CH$_2$C**H**$_2$CH$_2$NH$_a$), 1.53, 1.64 (2m, 4H, OCH$_2$C**H**$_2$CH$_2$CH$_2$NH$_a$), 2.83 (m, 4H, OC**H**$_2$CH$_2$CH$_2$-CH$_2$NH$_a$ inverted), 3.23 (m, 4H, OCH$_2$CH$_2$CH$_2$C**H**$_2$NH$_a$), 3.40, 3.56 (2m, 4H, OC**H**$_2$CH$_2$-CH$_2$CH$_2$NH$_a$), 4.14, 4.69 (ABq, 4H, $J = 12.1$ Hz, CH$_2$OCH$_2$), 4.20, 4.28 (Abq, 4H, $J = 11.0$ Hz, CH$_2$OCH$_2$), 4.44, 4.83 (Abq, 4H, $J = 11.4$ Hz, CH$_2$OCH$_2$), 4.99 (t, 1H, NH$_a$ inverted), 5.89 (t, 2H, NH$_a$), 7.25, 7.36 (2d, 4H, ArH), 7.32–7.36 (m, 4H, NH$_b$-inverted and Napht), 7.37 (s, 2H, ArH inverted), 7.40, 7.59 (2t, 4H, Napht), 7.50–7.54 (d + t, 4H, Napht), 7.68, 7.76, 7.82, 7.90, 7.96, 7.99, 8.20 (7d, 10H, Napht), 7.71 (s, 2H, NH$_b$); NMR ^{13}C (CDCl$_3$, 125.8 MHz) δ 25.0, 26.3 (OCH$_2$CH$_2$C**H**$_2$CH$_2$NH$_a$), 27.77, 27.80 (OCH$_2$C**H**$_2$CH$_2$CH$_2$NH$_a$), 31.4, 31.5 [C(**C**H$_3$)$_3$], 34.31, 34.33 [**C**(CH$_3$)$_3$], 40.0 (2C) (OCH$_2$CH$_2$CH$_2$C**H**$_2$NH$_a$), 64.0, 66.3, 68.4

(CH$_2$OCH$_2$), 74.8, 75.3 (OCH$_2$CH$_2$CH$_2$CH$_2$NH$_a$), 119.8, 120.9, 121.3, 124.40, 124.43, 125.8, 125.9 (3C), 126.0, 126.1, 126.2, 127.4, 127.6, 128.5, 128.7, 128.8 (ArH), 127.5, 127.6, 129.4, 130.1, 130.3, 133.9, 134.2 (2C), 134.4, 146.3, 146.6, 155.3 (Ar), 156.1, 157.2 (CO). Anal. Calcd for C$_{81}$H$_{96}$N$_6$O$_9$: C 74.97; H 7.46; N 6.48. Found: C 75.04; H 7.82; N 6.24.

*7,15,23-Tri-tert-butyl-25,26,27-tri[[(**N′**-1-naphthylthioureido)butyl]oxy]-2,3,10,11,18,19-hexahomo-3,11,19-trioxacalix[3]arene* (**4b**): Flash chromatography (SiO$_2$, eluent CH$_2$Cl$_2$/MeOH, from 99.9:0.1 to 99.3:0.7); it was obtained in 71% yield (0.64 g); m.p. 132–134 °C; ^{1}H NMR (CDCl$_3$, 500 MHz) δ 0.56 (m, 2H, OCH$_2$**CH$_2$**CH$_2$CH$_2$NH$_a$ inverted), 0.74 (m, 2H, OCH$_2$CH$_2$CH$_2$CH$_2$NH$_a$ inverted), 1.23 [s, 9H, C(CH$_3$)$_3$ inverted], 1.24 [s, 18H, C(CH$_3$)$_3$], 1.37 (m, 8H, OCH$_2$CH$_2$**CH$_2$**CH$_2$NH$_a$), 2.38 (m, 2H, O**CH$_2$**CH$_2$CH$_2$CH$_2$NH$_a$ inverted), 2.91 (m, 2H, OCH$_2$CH$_2$CH$_2$CH$_2$NH$_a$ inverted), 3.36 (m, 4H, O**CH$_2$**CH$_2$CH$_2$CH$_2$NH$_a$), 3.42 (m, 4H, OCH$_2$CH$_2$CH$_2$CH$_2$NH$_a$), 3.72, 4.36 (Abq, 4H, J = 10.9 Hz, CH$_2$OCH$_2$), 3.73, 3.99 (Abq, 4H, J = 11.0 Hz, CH$_2$OCH$_2$), 3.99, 4.44 (Abq, 4H, J = 10.1 Hz, CH$_2$OCH$_2$), 5.97 (broad t, 1H, NH$_a$ inverted), 6.41 (broad t, 2H, NH$_a$), 7.07 (broad s, 4H, ArH), 7.12 (s, 2H, ArH inverted), 7.43 (d, 2H, Napht), 7.47–7.52 (m, 8H, Napht), 7.57 (m, 2H, Napht), 7.73 (s, 2H, NH$_b$), 7.74 (s, 1H, NH$_b$ inverted), 7.83–7.88 (m, 5H, Napht), 7.90–7.94 (m, 3H, Napht), 8.05 (d, 1H, Napht); ^{13}C NMR (CDCl$_3$, 125.8 MHz) δ 24.2, 25.2 (OCH$_2$CH$_2$CH$_2$CH$_2$NH$_a$), 27.4, (OCH$_2$CH$_2$CH$_2$CH$_2$NH$_a$), 31.5, 31.6 [C(CH$_3$)$_3$], 34.2 [**C**(CH$_3$)$_3$], 44.7, 45.3 (OCH$_2$CH$_2$CH$_2$-CH$_2$NH$_a$), 64.2, 66.3, 68.5 (CH$_2$OCH$_2$), 73.4, 74.7 (OCH$_2$CH$_2$CH$_2$CH$_2$NH$_a$), 122.4, 122.6, 125.2, 125.3, 125.6, 125.7, 126.9, 127.0, 127.3, 127.38, 127.44, 127.8, 128.3, 128.56, 128.59, 128.62 (ArH), 129.5, 130.0, 130.2, 130.9, 132.3, 132.5, 134.6, 145.8, 145.9, 154.4, 155.2 (Ar), 181.6, 181.8 (CS). Anal. Calcd for C$_{81}$H$_{96}$N$_6$O$_6$S$_3$: C 72.29; H 7.19; N 6.24; S 7.15. Found: C 71.87; H 7.16; N 6.21; S 6.86.

*7,15,23-Tri-tert-butyl-25,26,27-tri[[(**N′**-1-pyrenylureido)butyl]oxy]-2,3,10,11,18,19-hexahomo-3,11,19-trioxacalix[3]arene* (**4c**): Recrystallization from CH$_2$Cl$_2$/*n*-hexane; it was obtained in 83% yield (0.85 g); m.p. 191–193 °C; IR (KBr) 3318 cm^{-1} (NH), 1638 cm^{-1} (CO); ^{1}H NMR (CDCl$_3$, 500 MHz) δ 0.74 (m, 4H, OCH$_2$**CH$_2$CH$_2$**CH$_2$NH$_a$ inverted), 1.26 [s, 18H, C(CH$_3$)$_3$], 1.36 [s, 9H, C(CH$_3$)$_3$ inverted], 1.48 (m, 4H, OCH$_2$CH$_2$**CH$_2$**CH$_2$NH$_a$), 1.57, 1.78 (2m, 4H, OCH$_2$**CH$_2$**CH$_2$CH$_2$NH$_a$), 2.84 (t, 2H, O**CH$_2$**CH$_2$CH$_2$CH$_2$NH$_a$ inverted), 2.92 (m, 2H, OCH$_2$CH$_2$CH$_2$C**H$_2$**NH$_a$ inverted), 3.19, 3.42 (2m, 4H, OCH$_2$CH$_2$CH$_2$C**H$_2$**NH$_a$), 3.46, 3.62 (2m, 4H, O**CH$_2$**CH$_2$CH$_2$CH$_2$NH$_a$), 4.15, 4.31 (Abq, 4H, J = 11.1 Hz, CH$_2$OCH$_2$), 4.19, 4.84 (Abq, 4H, J = 11.5 Hz, CH$_2$OCH$_2$), 4.49, 4.86 (Abq, 4H, J = 11.5 Hz, CH$_2$OCH$_2$), 5.03 (t, 1H, NH$_a$ inverted), 5.94 (t, 2H, NH$_a$), 7.26, 7.43 (2d, 4H, ArH), 7.42 (s, 2H, ArH inverted), 7.48, 7.60 (2d, 4H, Pyr), 7.67 (m, 3H Pyr), 7.73–7.80 (m, 4H, Pyr), 7.79 (s, 1H, NH$_b$ inverted), 7.85–7.92 (m, 4H, Pyr), 7.89 (s, 2H, NH$_b$), 7.98–8.06, 8.10–8.18 (m, 9H, Pyr), 8.22, 8.42, 8.55 (3d, 3H, Pyr); ^{13}C NMR (CDCl$_3$, 125.8 MHz) δ 25.0, 26.4 (OCH$_2$CH$_2$CH$_2$CH$_2$NH$_a$), 27.7, 27.8 (OCH$_2$CH$_2$CH$_2$CH$_2$NH$_a$), 31.4, 31.6 [C(CH$_3$)$_3$], 34.3, 34.4 [**C**(CH$_3$)$_3$], 40.1 (2C) (OCH$_2$CH$_2$CH$_2$**C**H$_2$NH$_a$), 64.0, 66.4, 68.5 (CH$_2$OCH$_2$), 74.7, 75.2 (OCH$_2$CH$_2$CH$_2$CH$_2$NH$_a$), 120.3, 120.5, 121.5, 121.6, 124.1, 124.6, 124.7, 124.8, 125.3, 125.56, 125.63, 125.7, 126.2, 126.4, 126.7, 126.9, 127.4, 127.5, 127.7, 127.8, 128.8 (ArH), 123.4, 124.2, 124.7, 125.0, 125.5, 127.9, 128.2, 129.4, 130.1, 130.3, 130.4, 131.0, 131.4, 131.5, 146.4, 146.6, 155.2 (Ar), 156.3, 157.5 (CO). Anal. Calcd for C$_{99}$H$_{102}$N$_6$O$_9$: C 78.23; H 6.76; N 5.53. Found: C 78.17; H 6.40; N 5.75.

3.3. ^{1}H NMR Titrations

The anion association constants (as log K_{ass}) were determined in CDCl$_3$ by ^{1}H NMR titration experiments. Several aliquots (up to 10 equiv.) of the anion solutions (as TBA salts) were added to 0.5 mL solution of the receptors (2.5 × 10^{-3} M) directly in the NMR tube. The spectra were recorded after each addition of the salts, and the temperature of the NMR probe was kept constant at 25 °C. The association constants were evaluated using the WinEQNMR2 program [34] and following the urea NH chemical shifts. The Job methods were performed keeping the total concentration in 2.5 × 10^{-3} M, the same concentration used in the ion-pair recognition studies.

3.4. UV–Vis Absorption and Fluorescence Studies

Absorption and fluorescence studies were conducted using a Shimadzu UV-3101PC UV-Vis–NIR spectrophotometer and a Fluorolog F112A fluorimeter in right-angle configuration, respectively. Association constants were determined in CH_2Cl_2 and MeCN by UV–Vis absorption spectrophotometry and by steady-state fluorescence at 25 °C. Absorption spectra were recorded between 250 and 430 nm, and the emission ones between 320 and 670 nm, and using quartz cells with an optical path length of 1 cm. Several aliquots (up to 10 equiv.) of the anion solutions (as TBA salts) were added to a 2 mL solution of the receptors ($1.0 \times 10^{-5} - 2.0 \times 10^{-5}$ M) directly in the cell. Emission spectra were corrected for the spectral response of the optics and the photomultiplier. Fluorescence quantum yields were measured using quinine sulphate as the reference ($\Phi_F = 0.546$ in H_2SO_4 0.5 M). Time-resolved fluorescence intensity decays were obtained using the single-photon timing method with laser excitation and microchannel plate detection, with the already described setup [38]. The excitation wavelengths used were at the maximum absorption of the calixarenes and the emission wavelengths at maximum emission, using a front-face geometry. The timescale varied from 12.2 to 24.4 ps per channel for derivative **4a**, and 12.2 to 136.7 ps per channel for derivative **4c**. The spectral changes were interpreted using the HypSpec 2014 program [39]. The thermodynamic parameters were obtained under the same conditions in a temperature range from 5 to 45 °C. Details concerning the photophysical properties determination were already provided [25].

3.5. Computational Details

Geometry optimisations and thermodynamics were calculated using the newly developed C++ library, ULYSSES [40]. The method of choice was GFN2 xTB [41] using ALPB for solvation effects [42]. The initial structures of the calixarene were generated with Avogadro [43,44]. Geometries were minimised using the BFGS algorithm along with the dogleg trust region with convergence criteria of 5×10^{-8} E_h for energies and 2.5×10^{-5} E_h/a_0 for gradients. The method of Lindh et al. was used to approximate the Hessian matrix in the step calculation [45]. As the intramolecular π-stacking interactions of naphthyl groups are extremely favorable in the gas phase, the resulting geometries were not faithful representations of the system in solution. Therefore, the structures were always optimised in the dielectric of acetonitrile. Further details are given in the Supporting Information.

4. Conclusions

The anion binding properties of three fluorescent (thio)ureido-hexahomotrioxacalix[3]arene receptors were investigated by NMR, UV–Vis absorption and fluorescence titrations. These derivatives, bearing naphthyl or pyrenyl moieties at the macrocycle lower rim linked by a butyl spacer, were obtained in the partial cone conformation in solution. Anions of different geometries were bound through hydrogen bonds in a 1:1 stoichiometry. The results showed that for all the receptors the association constants increase with the anion basicity and the carboxylates AcO^- and Bzo^-, and the halide F^- were the best bound anions. Pyrenyl urea **4c** is a more efficient receptor than naphthyl urea **4a**, as shown by all the spectroscopic methods used. This may be due to the presence of the bulkier pyrenyl moiety that makes **4c** less flexible and consequently more preorganised. Fluorescence of **4c** displays both monomer and excimer fluorescence. The thermodynamics data obtained in acetonitrile for **4a** with F^- and AcO^- anions indicated that the binding process is entropy-driven. The solvation of this type of macrocyclic skeleton, with three oxygen bridges (CH_2OCH_2), is expected to be stronger than that of a dihomooxacalix[4]arene, for example, and may explain the low enthalpy and the high entropy terms obtained [36]. Computational studies performed showed the critical role of the solvent on the anion-receptor association, approaching the computed enthalpy and entropy to the experimental values. As ditopic receptors, ureas **4a** and **4c** both showed some ion pair recognition, but at low temperature, the complexation/decomplexation process is still fast on the NMR time scale.

Supplementary Materials: The following are available online: https://www.mdpi.com/article/10.3 390/molecules27103247/s1. Titration curves with TBA salts in $CDCl_3$; Job's plots; absorption and emission spectra with TBA salts; computational details; ^1H, ^{13}C, DEPT and COSY NMR spectra. References [46–54] are cited in the supplementary.

Author Contributions: A.S.M.: investigation, data acquisition and analysis, and editing; P.M.M.: conceptualisation, supervision, acquisition, data analysis and interpretation, and writing, review, and editing; J.R.A.: NMR data analysis and interpretation, writing and review; M.N.B.-S.: photophysics data analysis, interpretation and writing; F.M.: theoretical calculations, analysis and writing. All authors have read and agreed to the published version of the manuscript.

Funding: Authors thank Fundação para a Ciência e a Tecnologia, projects UIDB/00100/2020 and UIDB/04565/2020. A. S. Miranda acknowledges PhD grant ref. SFRH/BD/129323/2017 and COVID/BD/152147/2022.

Institutional Review Board Statement: Not applicable.

Informed Consent Statement: Not applicable.

Data Availability Statement: Not applicable.

Acknowledgments: The authors thank Alexander Fedorov for performing the fluorescence lifetime measurements.

Conflicts of Interest: The authors declare no conflict of interest.

Sample Availability: Samples of the compounds are not available from the authors.

References

1. Kim, J.S.; Quang, D.T. Calixarene-derived fluorescent probes. *Chem. Rev.* **2007**, *107*, 3780–3799. [CrossRef]
2. Kumar, R.; Jung, Y.; Kim, J.S. Fluorescent calixarene hosts. In *Calixarenes and Beyond*; Neri, P., Sessler, J.L., Wang, M.-X., Eds.; Springer International Publishing: Urdorf, Switzerland, 2016; pp. 743–760.
3. Kumar, R.; Sharma, A.; Singh, H.; Suating, P.; Kim, H.S.; Sunwoo, K.; Shim, I.; Gibb, B.C.; Kim, J.S. Revisiting fluorescent calixarenes: From molecular sensors to smart materials. *Chem. Rev.* **2019**, *119*, 9657–9721. [CrossRef] [PubMed]
4. Gutsche, C.D. *Calixarenes, An Introduction*; Monographs in Supramolecular Chemistry; The Royal Society of Chemistry: Cambridge, UK, 2008.
5. Neri, P.; Sessler, J.L.; Wang, M.-X. (Eds.) *Calixarenes and Beyond*; Springer International Publishing: Urdorf, Switzerland, 2016.
6. Jeon, N.J.; Ryu, B.J.; Lee, B.H.; Nam, K.C. Fluorescent sensing of tetrahedral anions with a pyrene urea derivative of calix[4]arene chemosensor. *Bull. Korean Chem. Soc.* **2009**, *30*, 1675–1677.
7. Hung, H.C.; Chang, Y.Y.; Luo, L.; Hung, C.H.; Diau, E.W.G.; Chung, W.S. Different sensing modes of fluoride and acetate based on a calix[4]arene with 25,27-bistriazolylmethylpyrenylacetamides. *Photochem. Photobiol. Sci.* **2014**, *13*, 370–379. [CrossRef] [PubMed]
8. Sutariya, P.G.; Pandya, A.; Lodha, A.; Menon, S.K. A pyrenyl linked calix[4]arene fluorescence probe for recognition of ferric and phosphate ions. *RSC Adv.* **2014**, *4*, 34922–34926. [CrossRef]
9. Chawla, H.M.; Munjal, P. Evaluation of calix[4]arene tethered Schiff bases for anion recognition. *J. Lumin.* **2016**, *179*, 114–121. [CrossRef]
10. Nemati, M.; Hosseinzadeh, R.; Zadmard, R.; Mohadjerani, M. Highly selective colorimetric and fluorescent chemosensor for fluoride based on fluorenone armed calix[4]arene. *Sens. Actuators B* **2017**, *241*, 690–697. [CrossRef]
11. Uttam, B.; Kandi, R.; Hussain, M.A.; Rao, C.P. Fluorescent lower rim 1,3-dibenzooxadiazole conjugate of calix[4]arene in selective sensing of fluoride in solution and in biological cells using confocal microscopy. *J. Org. Chem.* **2018**, *83*, 11850–11859. [CrossRef]
12. Li, Z.Y.; Su, H.K.; Tong, H.X.; Yin, Y.; Xiao, T.; Sun, X.Q.; Jiang, J.; Wang, L. Calix[4]arene containing thiourea and coumarin functionality as highly selective fluorescent and colorimetric chemosensor for fluoride ion. *Spectrochim. Acta Part A Mol. Biomol. Spectrosc.* **2018**, *200*, 307–312. [CrossRef]
13. Nemati, M.; Hosseinzadeh, R.; Mohadjerani, M. Colorimetric and fluorimetric chemosensor based on upper rim-functionalized calix[4]arene for selective detection of fluoride ion. *Spectrochim. Acta Part A Mol. Biomol. Spectrosc.* **2021**, *245*, 118950. [CrossRef]
14. Capici, C.; De Zorzi, R.; Gargiulli, C.; Gattuso, G.; Geremia, S.; Notti, A.; Pappalardo, S.; Parisi, M.F.; Puntoriero, F. Calix[5]crown-3-based heteroditopic receptors for *n*-butylammonium halides. *Tetrahedron* **2010**, *66*, 4987–4993. [CrossRef]
15. Jeon, N.J.; Ryu, B.J.; Park, K.D.; Lee, Y.J.; Nam, K.C. Tetrahedral anions selective fluorescent calix[6]arene receptor containing urea and pyrene moieties. *Bull. Korean Chem. Soc.* **2010**, *31*, 3809–3811. [CrossRef]
16. Brunetti, E.; Picron, J.-F.; Flidrova, K.; Bruylants, G.; Bartik, K.; Jabin, I. Fluorescent chemosensors for anions and contact ion pairs with a cavity-based selectivity. *J. Org. Chem.* **2014**, *79*, 6179–6188. [CrossRef] [PubMed]
17. Evans, N.H.; Beer, P.D. Advances in anion supramolecular chemistry: From recognition to chemical applications. *Angew. Chem. Int. Ed.* **2014**, *53*, 11716–11754. [CrossRef] [PubMed]

18. Busschaert, N.; Caltagirone, C.; Van Rossom, W.; Gale, P.A. Applications of supramolecular anion recognition. *Chem. Rev.* **2015**, *115*, 8038–8155. [CrossRef]

19. Gale, P.A.; Howe, E.N.W.; Wu, X. Anion receptor chemistry. *Chem* **2016**, *1*, 351–422. [CrossRef]

20. He, Q.; Vargas-Zúñiga, G.I.; Kim, S.H.; Kim, S.K.; Sessler, J.L. Macrocycles as ion pair receptors. *Chem. Rev.* **2019**, *119*, 9753–9835. [CrossRef]

21. Cottet, K.; Marcos, P.M.; Cragg, P.J. Fifty years of oxacalix[3]arenes: A review. *Beilstein J. Org. Chem.* **2012**, *8*, 201–226. [CrossRef]

22. Marcos, P.M. Functionalization and properties of homooxacalixarenes. In *Calixarenes and Beyond*; Neri, P., Sessler, J.L., Wang, M.-X., Eds.; Springer International Publishing: Urdorf, Switzerland, 2016; pp. 445–466.

23. Teixeira, F.A.; Marcos, P.M.; Ascenso, J.R.; Brancatelli, G.; Hickey, N.; Geremia, S. Selective binding of spherical and linear anions by tetraphenyl(thio)urea-based dihomooxacalix[4]arene receptors. *J. Org. Chem.* **2017**, *82*, 11383–11390. [CrossRef]

24. Augusto, A.S.; Miranda, A.S.; Ascenso, J.R.; Miranda, M.Q.; Félix, V.; Brancatelli, G.; Hickey, N.; Geremia, S.; Marcos, P.M. Anion recognition by partial cone dihomooxacalix[4]arene-based receptors bearing urea groups: Remarkable affinity for benzoate ion. *Eur. J. Org. Chem.* **2018**, *2018*, 5657–5667. [CrossRef]

25. Miranda, A.S.; Serbetci, D.; Marcos, P.M.; Ascenso, J.R.; Berberan-Santos, M.N.; Hickey, N.; Geremia, S. Ditopic receptors based on dihomooxacalix[4]arenes bearing phenylurea moieties with electron-withdrawing groups for anions and organic ion pairs. *Front. Chem.* **2019**, *7*, 758. [CrossRef] [PubMed]

26. Miranda, A.S.; Martelo, L.M.; Fedorov, A.A.; Berberan-Santos, M.N.; Marcos, P.M. Fluorescence properties of *p-tert*-butyldihomooxacalix[4]arene derivatives and the effect of anion complexation. *N. J. Chem.* **2017**, *41*, 5967–5973. [CrossRef]

27. Miranda, A.S.; Marcos, P.M.; Ascenso, J.R.; Berberan-Santos, M.N.; Schurhammer, R.; Hickey, N.; Geremia, S. Dihomooxacalix[4]arene-based fluorescent receptors for anion and organic ion pair recognition. *Molecules* **2020**, *25*, 4708. [CrossRef] [PubMed]

28. Ni, X.; Zheng, X.; Redshaw, C.; Yamato, T. Ratiometric fluorescent receptors for both Zn^{2+} and $H_2PO_4^-$ ions based on a pyrenyl-linked triazole-modified homooxacalix[3]arene: A potential molecular traffic signal with an R-S latch logic circuit. *J. Org. Chem.* **2011**, *76*, 5696–5702. [CrossRef]

29. Wu, Y.; Ni, X.-L.; Mou, L.; Jin, C.-C.; Redshaw, C.; Yamato, T. Synthesis of a ditopic homooxacalix[3]arene for fluorescence enhanced detection of heavy and transition metal ions. *Supramol. Chem.* **2015**, *27*, 501–507. [CrossRef]

30. Wu, C.; Zhao, J.-L.; Jiang, X.-K.; Wang, C.-Z.; Ni, X.-L.; Zeng, X.; Redshaw, C.; Yamato, T. A novel fluorescence "on-off-on" chemosensor for Hg^{2+} via a water-assistant blocking heavy atom effect. *Dalton Trans.* **2016**, *45*, 14948–14953. [CrossRef]

31. Wu, C.; Wang, C.-Z.; Zhu, Q.; Zeng, X.; Redshaw, C.; Yamato, T. Click synthesis of a quinoline-functionalized hexahomotrioxa-calix[3]arene: A turn-on fluorescence chemosensor for Fe^{3+}. *Sens. Actuators B* **2018**, *254*, 52–58. [CrossRef]

32. Teixeira, F.A.; Ascenso, J.R.; Cragg, P.J.; Hickey, N.; Geremia, S.; Marcos, P.M. Recognition of anions, monoamine neurotransmitter and trace amine hydrochlorides by ureido-hexahomotrioxacalix[3]arene ditopic receptors. *Eur. J. Org. Chem.* **2020**, *13*, 1930–1940. [CrossRef]

33. Lambert, S.; Bartik, K.; Jabin, I. Specific binding of primary ammonium ions and lysine-containing peptides in protic solvents by hexahomotrioxacalix[3]arenes. *J. Org. Chem.* **2020**, *85*, 10062–10071. [CrossRef]

34. Hynes, M.J. EQNMR: A computer program for the calculation of stability constants from nuclear magnetic resonance chemical shift data. *J. Chem. Soc. Dalton Trans.* **1993**, 311–312. [CrossRef]

35. Bryantsev, V.S.; Hay, B.P. Conformational preferences and internal rotation in alkyl- and phenyl-substituted thiourea derivatives. *J. Phys. Chem. A* **2006**, *110*, 4678–4688. [CrossRef] [PubMed]

36. Marcos, P.M.; Ascenso, J.R.; Segurado, M.A.P.; Cragg, P.J.; Michel, S.; Hubscher-Bruder, V.; Arnaud-Neu, F. Lanthanide cation binding properties of homooxacalixarene diethylamide derivatives. *Supramol. Chem.* **2011**, *23*, 93–101. [CrossRef]

37. Appel, E.A.; Forster, R.A.; Koutsioubas, A.; Toprakcioglu, C.; Scherman, O.A. Activation energies control the macroscopic properties of physically cross-linked materials. *Angew. Chem. Int. Ed.* **2014**, *53*, 10038–10043. [CrossRef] [PubMed]

38. Menezes, F.; Fedorov, A.; Baleizao, C.; Valeur, B.; Berberan-Santos, M.N. Methods for the analysis of complex fluorescence decays: Sum of Becquerel functions versus sum of exponentials. *Methods Appl. Fluoresc.* **2013**, *1*, 015002. [CrossRef]

39. Gans, P.; Sabatini, A.; Vacca, A. Investigation of equilibria in solution. Determination of equilibrium constants with the HYPERQUAD suite of programs. *Talanta* **1996**, *43*, 1739–1753. [CrossRef]

40. Menezes, F.; Popowicz, G.M. ULYSSES: An efficient and easy to use semi-empirical library for C++. 2022, *Submitted Manuscript*.

41. Bannwarth, C.; Ehlert, S.; Grimme, S. GFN2-xTB—An accurate and broadly parametrized self-consistent tight-binding quantum chemical method with multipole electrostatics and density-dependent dispersion contributions. *J. Chem. Theory Comput.* **2019**, *15*, 1652–1671. [CrossRef]

42. Ehlert, S.; Stahn, M.; Spicher, S.; Grimme, S. Robust and efficient implicit solvation model for fast semiempirical methods. *J. Chem. Theory Comput.* **2021**, *17*, 4250–4261. [CrossRef]

43. Avogadro: An Open-Source Molecular Builder and Visualization Tool. Version 1.20. Available online: http://avogadro.cc/ (accessed on 10 April 2022).

44. Hanwell, M.D.; Curtis, D.E.; Lonie, D.C.; Vandermeersch, T.; Zurek, E.; Hutchison, G.R. Avogadro: An advanced semantic chemical editor, visualization, and analysis platform. *J. Cheminform.* **2012**, *4*, 17. [CrossRef]

45. Lindh, R.; Bernhardsson, A.; Karlström, G.; Malmqvist, P.-A. On the use of a Hessian model function in molecular geometry optimizations. *Chem. Phys. Lett.* **1995**, *241*, 423–428. [CrossRef]

46. Grimme, S. Supramolecular binding thermodynamics by dispersion-corrected density functional theory. *Chem. Eur. J.* **2012**, *18*, 9955–9964. [CrossRef]

47. Pracht, P.; Bohle, F.; Grimme, S. Automated exploration of the low-energy chemical space with fast quantum chemical methods. *Phys. Chem. Chem. Phys.* **2020**, *14*, 7169–7192. [CrossRef] [PubMed]

48. Spicher, S.; Grimme, S. Robust atomistic modeling of materials, organometallic, and biochemical systems. *Angew. Chem.* **2020**, *59*, 15665. [CrossRef] [PubMed]

49. Sure, R.; Grimme, S. Comprehensive benchmark of association (free) energies of realistic host–guest complexes. *J. Chem. Theory Comput.* **2015**, *11*, 3785–3801. [CrossRef] [PubMed]

50. Hunter, J.D. Matplotlib: A 2D graphics environment. *Comput. Sci. Eng.* **2007**, *9*, 90–95. [CrossRef]

51. Stewart, J.J.P. Optimization of parameters for semiempirical methods V: Modification of NDDO approximations and application to 70 elements. *J. Mol. Model.* **2007**, *13*, 1173–1213. [CrossRef]

52. Korth, M. Third-generation hydrogen-bonding corrections for semiempirical QM methods and force fields. *J. Chem. Theory Comput.* **2010**, *6*, 3808–3816. [CrossRef]

53. Kromann, J.C.; Christensen, A.S.; Steinmann, C.; Korth, M.; Jensen, J.H. A third-generation dispersion and third-generation hydrogen bonding corrected PM6 method: PM6-D3H+. *Peer J.* **2014**, *2*, 449. [CrossRef]

54. Grimme, S.; Antony, J.; Ehrlich, S.; Krieg, H. A consistent and accurate ab initio parametrization of density functional dispersion correction (DFT-D) for the 94 elements H-Pu. *J. Chem. Phys.* **2010**, *132*, 154104. [CrossRef]

 *molecules*

Article

Electronic Tuning of Host-Guest Interactions within the Cavities of Fluorophore-Appended Calix[4]arenes

Varun Rawat and Arkadi Vigalok *

School of Chemistry, The Raymond and Beverly Sackler Faculty of Exact Sciences, Tel Aviv University, Tel Aviv 69978, Israel
* Correspondence: avigal@tauex.tau.ac.il

Abstract: A series of fluorescent calix[4]arene scaffolds bearing electron-rich carbazole moiety conjugated at the lower rim have been prepared. Studies of the fluorescence quenching in the presence of the N-methyl pyridinium guest revealed that the electronic properties of the distal phenolic ring play a major role in the host–guest complexation. In particular, placing an electron-donating piperidine fragment at that ring significantly increased the host–guest interactions, while introducing the same fragment into the proximal phenolic ring weakened the fluorescence response. These results suggest that the dominant interactions between the guest and calixarene cavity involve the oxygen-depleted fluorophore-bearing aromatic ring and not the more electron-rich unsubstituted phenolic fragments.

Keywords: calixarene; chemosensors; fluorescence; molecular recognition; π interactions

Citation: Rawat, V.; Vigalok, A. Electronic Tuning of Host-Guest Interactions within the Cavities of Fluorophore-Appended Calix[4]arenes. *Molecules* **2022**, *27*, 5689. https://doi.org/10.3390/molecules27175689

Academic Editor: Paula M. Marcos

Received: 2 August 2022
Accepted: 2 September 2022
Published: 3 September 2022

Publisher's Note: MDPI stays neutral with regard to jurisdictional claims in published maps and institutional affiliations.

1. Introduction

Complexation of cationic organic guests within the electron-rich cavity of calix[4]arene (calixarene) compounds has been at the heart of the host–guest complexation chemistry for several decades [1]. In particular, multiple calixarene scaffolds have been investigated in much detail with regard to the complexation of various pyridinium salts and their derivatives [2]. In addition to common calixarene hosts, these studies involved calixarene scaffolds adapting various conformations [3], and scaffolds containing two calixarene cavities (Figure 1) [4]. In the great majority of the studies, ^{1}H NMR spectroscopy was the method of choice to determine the strength of the host–guest complexation [5], with the technique typically requiring relatively high concentrations. Surprisingly, to our knowledge, studies on common electronic effects on this complexation reaction have not been reported. While the introduction of electron-donating or -accepting substituents in the calixarene aromatic rings can be viewed as a judicious, albeit synthetically challenging, route to study these effects, the data analysis can be skewed by the conformational changes of the host molecule with regard to the guest cation. It is generally accepted that π interactions (cation-π and/or π-stacking) play an important role in the overall complexation of N-alkyl pyridinium salts within the calixarene hosts [6]. With nearly all studied calix[4]arene hosts having four alkyl ether groups at the lower rim, the average C4v conic structure of the cavity is not optimized for such interactions [7–16]. Naturally, π interactions would be maximized if a pair of the opposite aromatic groups adopted a parallel disposition, where two opposite aromatic groups are parallel to each other in a C_{2v} symmetrical conformation (**1d**, flattened cone), which for the symmetrically substituted calixarenes can be observed only at low temperatures [16–18]. A straightforward way to achieve such an arrangement is the selective 1,3-lower rim dialkylation or acylation of the phenolic oxygens, leading to the protected phenolic moieties adopting a parallel geometry. Alternatively, a replacement of an oxygen atom at the lower rim with a non-polar hydrocarbyl group also results in the oxygen-depleted (formerly) phenolic fragment becoming aligned with the opposite phenolic ring, as can be deduced from the available structural data (**1e**) [19,20]. Interestingly, although these parallel aromatic rings are pre-arranged to participate in π interactions, they are less electron rich than the remaining unsubstituted phenolic rings which can adopt a similar arrangement by sacrificing the stabilizing hydrogen bonding between the OH groups. Distinguishing between the two different binding modes could be aided by studying electronic effects of appropriate substituents on the cation complexation. Surprisingly, no such studies have been reported to the best of our knowledge.

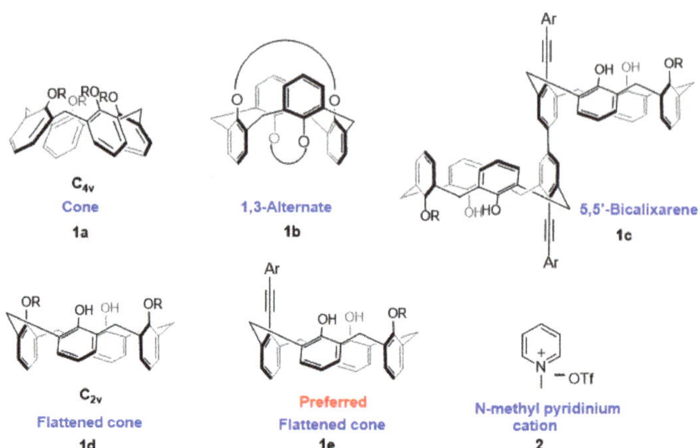

Figure 1. Representative calixarene scaffolds for host–guest chemistry.

While studying the chemosensory properties of oxygen-depleted 5,5′-Bicalixarene scaffolds (**1c**) bearing an alkyne function at the lower rim [21], we discovered that these compounds show strong NMR and fluorescence response upon the complexation of N-methyl pyridinium cation (**2**) [20,22]. Attachment of electron-donating fluorophores at the termini of the bicalixarene fragment expectedly increased the host–guest complexation properties of the scaffolds [23]. Yet, this observation alone does not provide compelling evidence for the π interactions with the oxygen-depleted part of the calixarene moiety. Here, we present our studies of model calixarene compounds that support the notion of the π interactions between the parallel opposing aromatic rings and N-methyl pyridinium cation playing major role in the host–guest complexation (Figure 2).

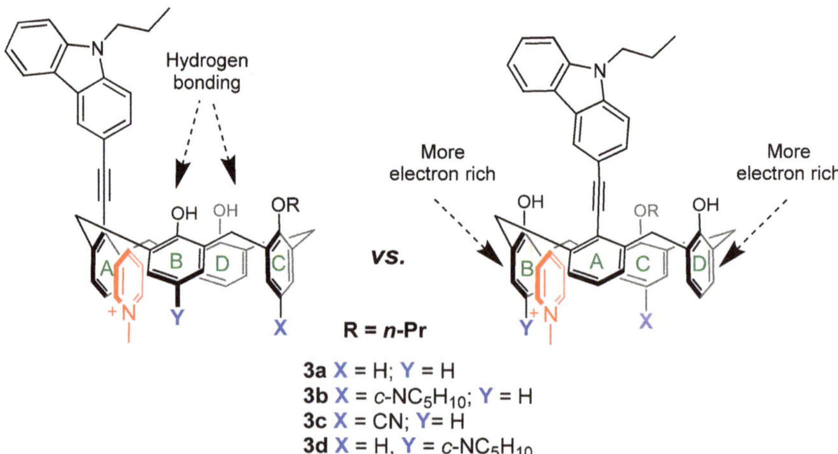

Figure 2. Proposed π interactions between **2** and carbazole-appended calixarenes **3**.

2. Results and Discussion

Although the presence of the electron-donating fluorophores at the termini of the biphenyl chain in **1c** increases the fluorescence response to the host–guest interactions with **2**, there is no evidence for this chain being involved in the π interactions. Because the adjacent free phenolic rings are more electron rich, they potentially can provide stronger π stabilization to the cationic aromatic guest. As stated above, such strong stabilization would come at the cost of breaking hydrogen bonding between the phenolic groups at the lower rim. To establish the pair of the opposing aromatic rings being responsible for the π interactions with **2**, we decided to directly compare its complexation within the cavities of the substituted mono calixarene hosts **3** (Figure 2).

We hypothesized that if the fluorophore-appended ring A is involved in the π interactions, the substituents in ring C should have major effect on the complexation of **2**. On the other hand, if the phenolic rings B and D are the main contributors to the π interactions, substitution in ring B will cause some change in the fluorescence response. To verify this hypothesis, we developed synthetic protocols toward unsymmetrically substituted hosts **3a–d** (Schemes 1–3). Moreover, although we earlier reported the synthesis of the parent compound **3a** (Φ = 0.17) [23], we have now modified the procedure to obtain the desired compound in only three steps (Scheme 1).

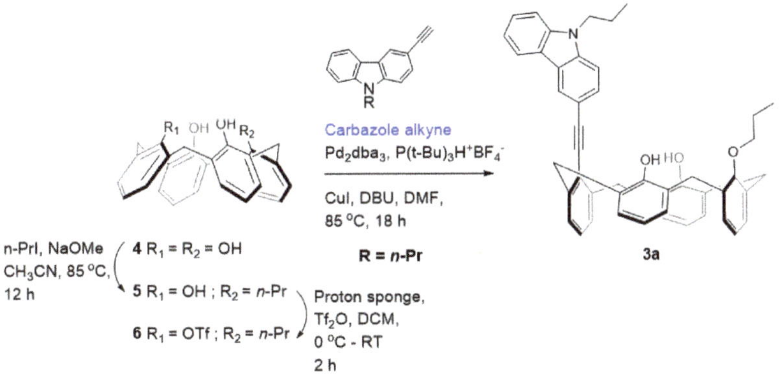

Scheme 1. Synthesis of compound **3a**.

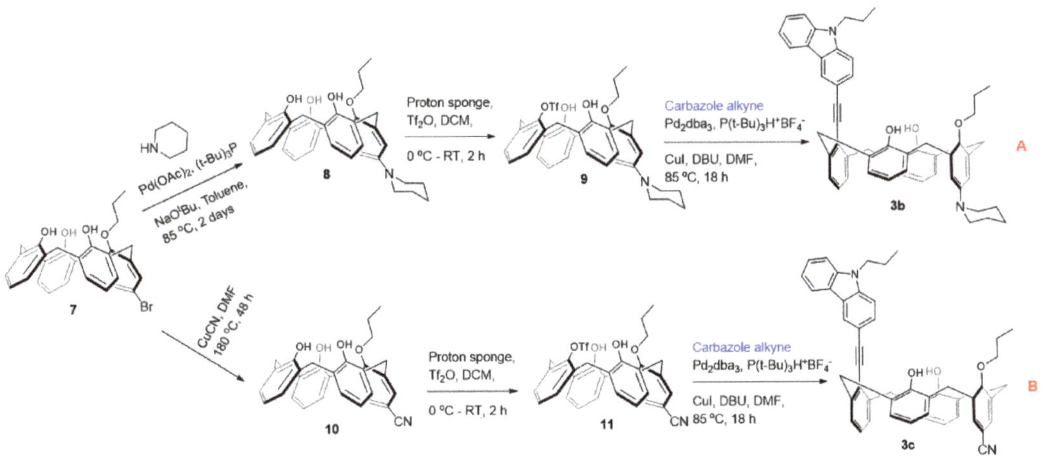

Scheme 2. Synthesis of compounds **3b,c**.

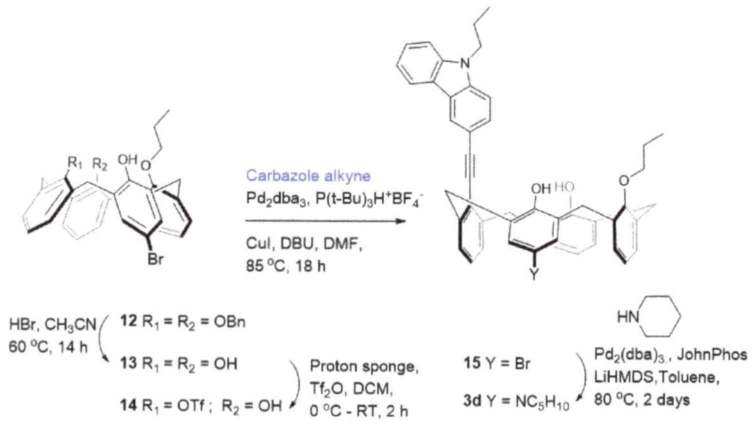

Scheme 3. Synthesis of compound **3d**.

To prepare compounds **3b** and **3c**, bearing, at the C ring, the electron-donating piperidine group and electron-withdrawing cyano group, respectively, the corresponding bromoderivative **7** was prepared in three steps [23]. Reacting compound **7** with piperidine under the Buchwald–Hartwig amination conditions afforded the amino derivative **8** which was converted to the triflate **9**. The Sonogashira coupling with the carbazole alkyne gave **3b** in a 14% overall yield and quantum yield of 30% ($\Phi = 0.30$) (Scheme 2A) [24].

For **3c** ($\Phi = 0.14$), compound **7** was converted to the cyano derivative **10** via the Rosenmund–von Braun reaction with CuCN, followed by the similar protocols for the installation of the carbazole group at the lower rim (Scheme 2B). To prepare compound **3d**, the selective protection of the phenolic groups on rings A, D, and C was performed followed by the bromination at the para- to the OH position of the remaining unprotected ring B (compound **12**) [25,26]. The removal of the benzylic groups and selective triflation of the intermediate **13** produced the triflate **14** which was reacted with **4**. Finally, the obtained compound **15** was converted to **3d** ($\Phi = 0.39$) via Buchwald–Hartwig amination with piperidine (Scheme 3) [27]. All new compounds were fully characterized by the multinuclear NMR spectroscopy and HRMS. All compounds **3a–d** exhibit strong fluorescence upon irradiation with the UV light (Figure 3).

With these fluorescent calixarenes in hand, we moved to explore their complexation properties toward **2**. As expected, addition of **2** to a 10 µM solution of a calixarene in 1,2-dichloroethane (DCE) resulted in the fluorescence decrease (Figure 4). Titration of the solutions of **3** with 1–10 equiv. of **2** allowed measurements of binding constants (Table 1), which were in the same range reported for calix[4]arene receptors from the UV measurements in chloroform at similar concentrations [10]. The overall numbers ($\sim$4000–6000 M^{-1}) are higher than K_{ass} obtained by the ^{1}H NMR technique ($K_{ass} = 162 \pm 13$ M^{-1} for **3c**) at higher concentrations. The latter compares well with the literature data for NMP cation complexation obtained by the ^{1}H NMR technique for calixarene hosts with a single cavity [9,10]. Importantly, the most significant drop in the fluorescence intensity was observed for compound **3b**, bearing an electron-rich piperidine moiety at the ring C opposite to the fluorophore unit. Calixarene **3a**, unsubstituted at the upper rim, showed a weaker response while **3c**, possessing an electron withdrawing cyano group, was the least responsive among these three compounds. Interestingly, calixarene **3d** showed the weakest response to the presence of the cation **2** despite having an electron-donating substituent at the upper rim (Figure 5, Table 1). These results suggest that the complexation of **2** within the calixarene cavity is directed by the π interactions with the aromatic rings A and C. With the electron-donating piperidine at ring C, the complexation is enhanced, while with the electron-withdrawing CN at ring C, the complexation is weakened compared with the parent **3a**. On the other hand, an electron-donating piperidine unit at ring B weakens cation complexation presumably due to repulsive steric interactions between the piperidine and **2** (Supplementary Materials, Figure S3). Thus, the preset parallel alignment of rings A and C appears more important in the cation complexation within the calixarene cavity over higher electron density in rings B and D, which are prevented from maximizing their π interactions due to hydrogen bonding at the lower rim.

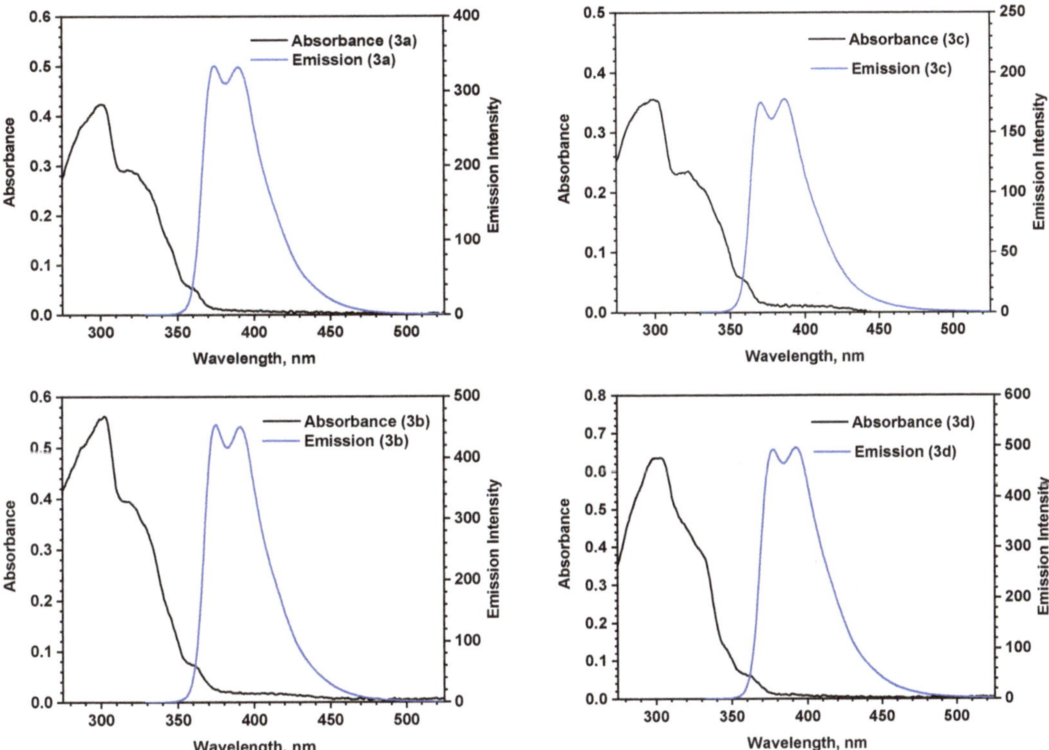

Figure 3. Absorbance–emission spectra of compounds **3a–d**.

Table 1. Emission intensity dependence on the concentration of **2** [a]. % Decrease in Emission Intensity.

Calixarene	1 Equiv. 2	5 Equiv. 2	10 Equiv. 2	K_{ass} [b]
3a	7%	17%	33%	$4392 \pm 150\,M^{-1}$
3b	9%	25%	39%	$5935 \pm 210\,M^{-1}$
3c [c]	6%	17%	30%	$3900 \pm 134\,M^{-1}$
3d	4%	8%	12%	$1182 \pm 76\,M^{-1}$

[a] Solutions of calixarenes **3a–d** in DCE (10 μM) were treated with increasing concentrations of NMPT, **2** (0 to 10 equivalents) at 24 °C; [b] the binding constants were calculated directly from the Stern–Volmer plots, see Supplementary Materials for details (page S43); [c] binding constant calculated via NMR measurements was found to be $162 \pm 13\,M^{-1}$, see Supplementary Materials for details (page S45).

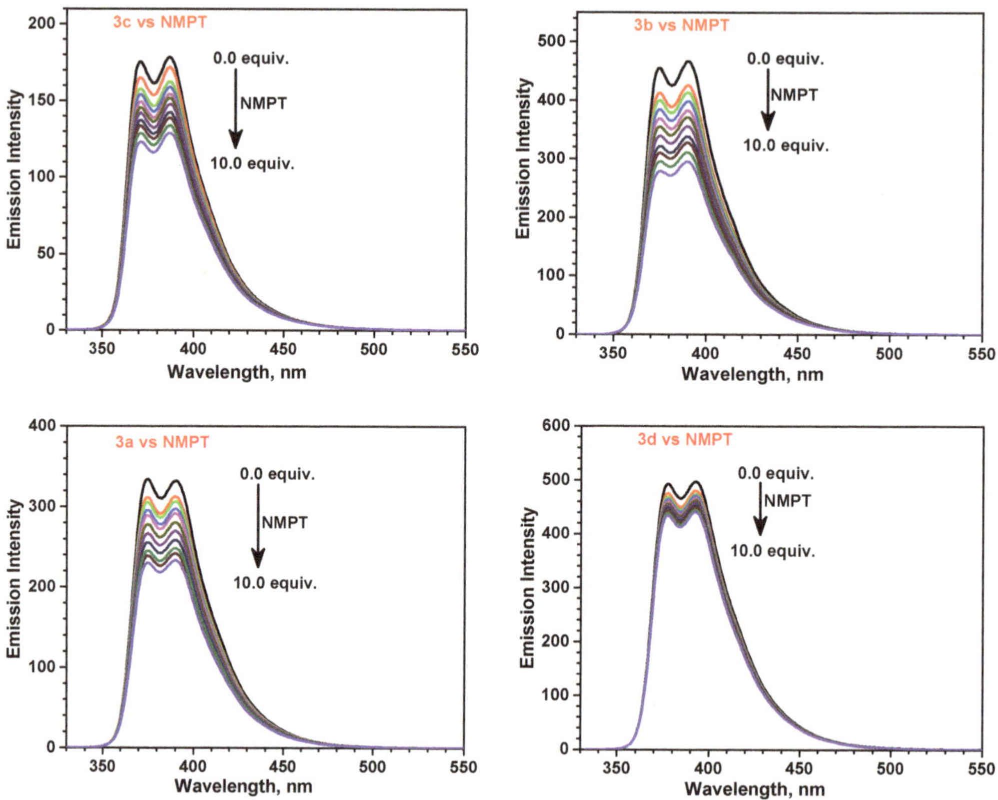

Figure 4. Fluorescent spectra of carbazole-appended calixarenes **3a–d** (DCE, 10 μM) with various concentrations of NMPT, **2** (0 to 10 equivalents). Excitation wavelength 321 nm.

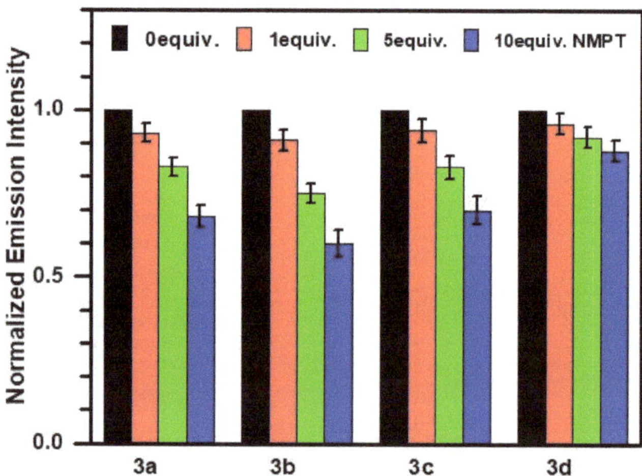

Figure 5. Fluorescence quenching in carbazole-appended calixarenes **3a–d** (DCE, 10 μM) upon the addition of N-methylpyridinium triflate (NMPT, **2**).

The fluorescence analysis was further corroborated by the ^{1}H NMR studies of complexation of **2** by **3a–d**. Unlike the fluorescence quenching which would likely depend on the guest orientation within the cavity, the chemical shifts of the host's protons should only reflect the strength of the host–guest interactions. At 5 mM concentrations, the 1:1 mixtures of **2** with **3a** or **3b** showed significant upfield shift for the N-CH$_3$ group and aromatic protons (Figure 6, Table 2). On the other hand, the same signals were only slightly shifted in the case of **3c** and **3d**, testifying to weaker host–guest interactions. Higher sensitivity of the aromatic protons in **2** suggests that the aromatic ring is likely partly immersed into the calixarene cavity with ensuing π–π interactions [10].

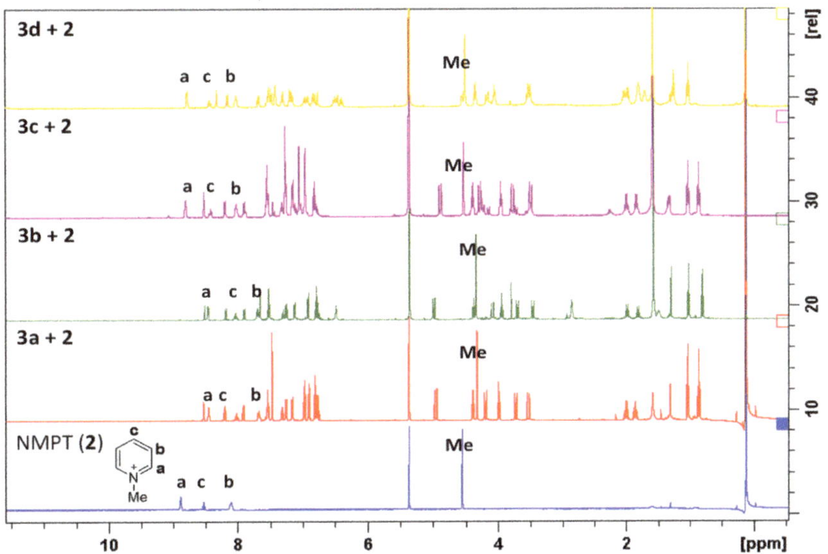

Figure 6. ^{1}H NMR (400 MHz) host–guest complexation studies in CD$_2$Cl$_2$ (5 mM).

Table 2. Comparative studies of the ^{1}H NMR chemical shifts (δ, ppm) of **2** within the calixarene cavities (CD$_2$Cl$_2$, 5 mM).

Resonance Signal	2	3a + 2	3b + 2	3c + 2	3d + 2
Me	4.56	4.33	4.36	4.53	4.51
H$_a$	8.88	8.45	8.47	8.80	8.78
H$_b$	8.11	7.68	7.71	8.02	8.02
H$_c$	8.54	8.02	8.05	8.42	8.44

3. Materials and Methods

The synthetic manipulations involving air-sensitive compounds were performed in a nitrogen-filled Innovative Technology or Vigor glove box. All solvents were degassed and stored under high-purity nitrogen and activated 4Å molecular sieves. All deuterated solvents were stored under high-purity nitrogen on 3Å molecular sieves. Commercially available reagents (Aldrich, Strem, and Fluka) were used as received. The NMR spectra were recorded on a Bruker Avance 400 MHz spectrometer. ^{1}H and ^{13}C NMR signals are reported in ppm downfield from TMS. All measurements were performed at 22 °C in CDCl$_3$/CD$_2$Cl$_2$ unless stated otherwise. Mass spectra were recorded on a VG-Autospec M-250 instrument. UV and fluorescence spectra were recorded on a Vernier fluorescence/UV-Vis spectrophotometer and Hitachi F-2710 fluorescence spectrophotometer.

Synthesis of 5: A sample of 4.24 g (10.0 mmol) of calix[4]arene **4** and 0.64 g (11.8 mmol) of NaOCH$_3$ was refluxed in 300 mL of CH$_3$CN for 30 min to monodeprotonate the calix[4]arene completely. To this, 2.4 mL (4.18 g, 24.6 mmol) of n-propyl iodide was added, and the reaction mixture was further refluxed for 12 h. After the completion of the reaction (monitored by TLC), the reaction mixture was neutralized with a few drops of acetic acid, and the solvent was removed to leave an off-white residue. The residue was dissolved in 150 mL $CHCl_3$ and successively washed with H_2O and brine. The organic phase was separated, dried over anhydrous $MgSO_4$, and evaporated. The residue was recrystallized from $CHCl_3$ with a slow addition of CH_3OH to yield the corresponding monoalkylated calix[4]arene 5. Yield: 3.26 g (70%); White solid; ^{1}H NMR (CDCl$_3$, 400 MHz): δ 10.25 (s, 1H), 9.79 (s, 1H), 9.48 (s, 1H), 7.05–7.13 (m, 6H), 6.93 (t, J = 7.7 Hz, 1H), 6.79 (t, J = 7.0 Hz, 3H), 6.17–6.76 (m, 2H), 4.45 (d, J = 13.4 Hz, 2H), 4.32 (d, J = 13.6 Hz, 2H), 4.18 (t, J = 6.9 Hz, 2H), 3.53 (d, J = 1.7 Hz, 2H), 3.50 (br s, 2H), 2.24 (q, J = 7.3, 14.7 Hz, 2H), 1.34 (t, J = 7.4 Hz, 3H). ^{13}C NMR (CDCl$_3$, 100 MHz): δ 151.58, 150.95, 149.36, 148.91, 134.39, 129.46, 129.11, 128.94, 128.88, 128.56, 128.37, 126.21, 122.38, 122.09, 121.05, 79.16, 32.05, 31.84, 31.56, 23.42, 10.80. ESI-MS calcd for [M+Na]$^+$ $C_{31}H_{30}NaO_4$ 489.20, found 489.46.

Synthesis of 6: To a suspension of 1.81 g (3.9 mmol) of mono-propyl ether 5 and 1,8-bis(dimethylamino)naphthalene (proton sponge) (2.16 g, 10.1 mmol) in dry CH_2Cl_2 (40 mL) at 0 °C, trifluoromethanesulphonic anhydride (1.3 mL, 7.8 mmol) was added under nitrogen. After 2 h of stirring at room temperature, the organic layer was washed twice with HCl 10% and once with water, dried over $MgSO_4$, and evaporated. The residue was subjected to column chromatography purification (CH_2Cl_2/Hexane 3/10 v/v) giving product 6. Yield: 1.82 g (78%); White solid; ^{1}H NMR (CDCl$_3$, 400 MHz): δ 7.34 (s, 2H), 7.17 (dd, J = 1.4, 7.5 Hz, 2H), 7.09 (dd, J = 1.5, 7.6 Hz, 2H), 7.05 (s, 1H), 7.02 (s, 1H), 6.88–6.92 (m, 3H), 6.79–6.84 (m, 1H), 6.75 (t, J = 7.6 Hz, 2H), 4.50 (d, J = 14.6 Hz, 2H), 4.19 (t, J = 5.5 Hz, 2H), 4.01 (d, J = 13.8 Hz, 2H), 3.58 (d, J = 13.8 Hz, 2H), 3.46 (d, J = 13.8 Hz, 2H), 2.19 (sextet, J = 7.7, 14.2 Hz, 2H), 1.34 (t, J = 7.5 Hz, 3H). ^{13}C NMR (CDCl$_3$, 100 MHz): δ 152.96, 149.95, 143.37, 133.99, 132.66, 129.96, 129.65, 129.07, 128.89, 128.63, 127.53, 127.07, 125.68, 119.63, 80.26, 31.95, 31.72, 23.19, 10.56. 19F NMR: −74.34 (s). ESI-MS calcd for [M+Na]$^+$ $C_{32}H_{29}F_3NaO_6S$ 621.15, found 621.41.

Synthesis of carbazole-appended calix[4]arene 3a: To a mixture of Pd$_2$dba$_3$ (0.05 equiv.) and P(t-Bu)$_3$H$^+$ BF$_4$$^-$ (0.2 equiv.) dissolved in 10 mL of dry DMF, CuI (2.5 equiv.), DBU (4 equiv.), carbazole alkyne (5 equiv.) and triflate 6 (0.25 mmol) were added and the mixture was heated at 85 °C in an oil bath for 12 h. The solvent was evaporated, and the resulting crude product was dissolved in CH_2Cl_2 and washed with brine several times. Drying the CH_2Cl_2 extract over $MgSO_4$ followed by solvent removal under vacuum gave the crude product. The residue was subjected to column chromatography (CH_2Cl_2/Hexane 4/10 v/v) to obtain the pure compound 3a. Yield: 0.106 g (62%); White solid; ^{1}H NMR (CD$_2$Cl$_2$, 400 MHz): δ 8.53 (d, J = 0.9 Hz, 1H), 8.19 (d, J = 7.8 Hz, 1H), 7.90 (dd, J = 1.6, 8.5 Hz, 1H), 7.52–7.54 (m, 3H), 7.36 (s, 2H), 7.29–7.33 (m, 1H), 7.22 (dd, J = 1.4, 7.5 Hz, 2H), 7.12 (dd, J = 1.5, 7.6 Hz, 2H), 7.02–7.04 (m, 2H), 6.87–6.96 (m, 4H), 6.76 (t, J = 7.5 Hz, 2H), 4.93 (d, J = 12.1 Hz, 2H), 4.39 (t, J = 8.0 Hz, 2H), 4.17 (d, J = 14.7 Hz, 2H), 3.99 (t, J = 5.3 Hz, 2H), 3.69 (d, J = 13.3 Hz, 2H), 3.51 (d, J = 13.3 Hz, 2H), 1.96–2.05 (m, 2H), 1.83–1.92 (m, 2H), 1.04 (t, J = 7.0 Hz, 3H), 0.87 (t, J = 7.9 Hz, 3H). ^{13}C NMR (CD$_2$Cl$_2$, 100 MHz): δ 153.8, 151.5, 141.7, 141.4, 140.4, 133.2, 129.8, 129.7, 129.5, 129.4, 129.1, 128.5, 128.4, 128.0, 127.7, 127.4, 127.2, 126.1, 125.8, 124.1, 123.0, 122.8, 120.8, 119.4, 114.3, 109.1, 108.8, 98.6, 86.5, 78.7, 44.9, 36.6, 31.9, 23.4, 22.5, 11.9, 10.7. HRMS (ESI-TOF) m/z [M+H]$^+$ calcd for $C_{48}H_{44}NO_3$ 682.3321, found 682.3323.

Synthesis of 8: The reaction was carried out under an inert atmosphere of pure nitrogen. To a stirred suspension of Pd(OAc)$_2$ (0.014 g, 0.063 mmol), P(t-Bu)$_3$ (0.019 g, 0.095 mmol) and sodium tert-butoxide (0.121 g, 1.26 mmol), in toluene (15 mL) were added piperidine (0.065 g, 0.75 mmol) and compound 7 (0.343 g, 0.63 mmol). The reaction mixture was then stirred at 85 °C for 48 h. The solvent was evaporated, and the resulting crude product was dissolved in EtOAc (50 mL), washed with water (5 mL × 2), brine, and dried over anhydrous $MgSO_4$. Removal of solvent under reduced pressure and column chromatographic purification with EtOAc/Hexane (2:8 v/v) gave pure compounds 8 in 78%

yields (0.270 g). Off-white solid: ^{1}H NMR (CD$_2$Cl$_2$, 400 MHz): δ 9.43 (br s, 2H), 6.99–7.09 (m, 7H), 6.63–6.71 (m, 5H), 4.34 (d, J = 12.6 Hz, 2H), 4.26 (d, J = 12.6 Hz, 2H), 4.08 (t, J = 6.1 Hz, 2H), 3.49 (d, J 12.6 Hz, 2H), 3.39 (d, J = 12.6 Hz, 2H), 3.93–3.96 (m, 4H), 2.16–2.19 (m, 4H), 1.61–1.63 (m 4H), 1.49–1.52 (m, 2H), 1.29 (t, J = 6.8 Hz, 3H). ^{13}C NMR (CD$_2$Cl$_2$, 100 MHz): δ 186.42, 177.34, 160.20, 158.22, 151.17, 150.30, 149.44, 144.46, 134.36, 133.99, 130.69, 129.22, 128.85, 128.76, 128.64, 128.52, 128.36, 128.15, 127.26, 126.01, 124.19, 121.90, 120.86, 120.25, 117.74, 117.41, 117.11, 79.18, 78.60, 78.34, 51.21, 51.04, 32.70, 31.87, 30.70, 30.55, 30.43, 26.16, 24.29, 23.43, 10.68. ESI-MS calcd for [M+H]$^+$ C$_{36}$H$_{40}$NO$_4$ 550.30, found 550.62.

Synthesis of 9: To a suspension of 8 (0.270 g, 0.49 mmol) and 1,8-bis(dimethylamino) naphthalene (proton sponge) (0.272 g, 1.27 mmol) in dry CH$_2$Cl$_2$ (20 mL) at 0 °C trifluoromethanesulfonic anhydride (0.276 g, 0.16 mL, 0.98 mmol) was added under nitrogen. After 2 h of stirring at room temperature, the organic layer was washed once with 10% HCl and once with water, dried over anhydrous MgSO$_4$, and evaporated. The residue was purified by column chromatography (silica gel, CH$_2$Cl$_2$/Hexane 4/10 v/v) to give the title compound as white solid. Yield: 0.244 g (73%); White solid; ^{1}H NMR (CD$_2$Cl$_2$, 400 MHz): δ 7.65 (br s, 2H), 7.19 (dd, J = 1.5, 7.6 Hz, 2H), 7.14 (dd, J = 1.5, 7.6 Hz, 2H), 7.02 (s, 1H), 6.99 (s, 1H), 6.92–6.94 (m, 1H), 6.75 (t, J = 7.5Hz, 2H), 6.62 (s, 2H), 4.50 (d, J = 12.8 Hz, 2H), 4.17 (t, J = 6.4 Hz, 2H), 3.99 (d, J = 12.8 Hz, 2H), 3.55 (d, J = 12.8 Hz, 2H), 3.48 (d, J = 12.8 Hz, 2H), 2.99 (t, J = 5.3 Hz, 2H), 2.15–2.36 (m, 2H), 1.51–1.67 (m, 4H), 1.52–1.56 (m, 2H), 1.34 (t, J = 7.5 Hz, 3H). ^{13}C NMR (CD$_2$Cl$_2$, 100 MHz): δ 153.03, 150.76, 143.28, 142.51, 134.13, 132.94, 130.04, 129.82, 128.88, 128.79, 128.58, 127.44, 125.93, 120.64, 119.70, 117.47, 117.21, 80.43, 50.44, 32.11, 31.86, 26.02, 24.18, 23.17, 10.36. 19F NMR: −74.74 (s). ESI-MS calcd for [M+H]$^+$ C$_{37}$H$_{39}$F$_3$NO$_6$S 682.25, found 682.47.

Synthesis of carbazole-appended calix[4]arene 3b: To a mixture of Pd$_2$dba$_3$ (0.05 equiv.) and P(t-Bu)$_3$H$^+$ BF$_4^-$ (0.2 equiv.) dissolved in 10 mL of dry DMF, CuI (2.5 equiv.), DBU (4 equiv.), carbazole alkyne (5 equiv.) and triflate 9 (0.25 mmol) were added and the mixture was heated at 85 °C in an oil bath for 12 h. The solvent was evaporated, and the resulting crude product was dissolved in CH$_2$Cl$_2$ and washed with brine several times. Drying the CH$_2$Cl$_2$ extract over anhydrous MgSO$_4$ followed by solvent removal under vacuum gave the crude product. The residue was subjected to column chromatography (CH$_2$Cl$_2$/Hexane 4/10 v/v) to obtain the pure compound. Yield: 0.090 g (47%); Slightly yellow solid; ^{1}H NMR (CD$_2$Cl$_2$, 400 MHz): δ 8.52 (d, J = 1.0 Hz, 1H), 8.20 (dd, J = 4.8, 3.9 Hz, 1H), 7.91 (dd, J = 8.5, 1.6 Hz, 1H), 7.62–7.49 (m, 5H), 7.28–7.34 (m, 1H), 7.22 (dd, J = 7.5, 1.5 Hz, 2H), 7.09 (dd, J = 7.6, 1.6 Hz, 2H), 7.02–6.90 (m, 3H), 6.71 (t, J = 7.5 Hz, 2H), 6.58 (s, 2H), 4.98 (d, J = 12.8 Hz, 2H), 4.39 (t, J = 7.1 Hz, 2H), 4.08 (d, J = 13.5 Hz, 2H), 3.96 (t, J = 6.6 Hz, 2H), 3.69 (d, J = 12.8 Hz, 2H), 3.46 (d, J = 12.9 Hz, 2H), 3.00–2.95 (m, 4H), 2.06–1.95 (m, 2H), 1.90–1.77 (m, 2H), 1.70–1.59 (m, 4H), 1.52–1.54 (m, 2H), 1.05 (t, J = 7.5 Hz, 3H), 0.82 (t, J = 7.4 Hz, 3H). ^{13}C NMR (CD$_2$Cl$_2$, 100 MHz): δ 153.64, 150.28, 143.67, 141.78, 141.10, 140.32, 133.32, 129.53, 129.03, 128.73, 128.51, 128.20, 127.83, 127.23, 127.09, 126.81, 126.09, 123.72, 123.32, 122.90, 122.65, 120.62, 120.18, 119.52, 119.31, 117.28, 114.75, 109.20, 108.97, 97.33, 87.30, 79.50, 50.70, 44.87, 36.33, 32.17, 26.14, 24.26, 23.33, 22.46, 11.64, 10.35. HRMS (ESI-TOF) m/z [M+H]$^+$ calcd for C$_{53}$H$_{53}$N$_2$O$_3$ 765.4056, found 765.4055.

Synthesis of 10: To a solution of 7 (1.0 g, 1.83 mmol) in DMF (50 mL), CuCN (0.492 g, 5.49 mmol) was added. The resulting heterogeneous mixture was poured into a thick wall glass pressure round bottom flask and then heated at 180 °C for 48 h under vigorous stirring. After cooling, the solvent was completely evaporated under reduced pressure. The resulting sticky residue was extracted twice with hot ethyl acetate (2 × 100 mL). The combined organic phases were then washed twice with brine (2 × 100 mL), dried over anhydrous MgSO$_4$, and then evaporated to dryness (the separated water phase was carefully treated with a solution of sodium hypochlorite to destroy the residuals cyanide ions). Purification of the solid residue by silica column chromatography (CH$_2$Cl$_2$/hexane, v/v 8:2) gave title compound. Yield: 0.603 g (67%); Slightly yellow solid; ^{1}H NMR (CDCl$_3$, 400 MHz): δ 9.43 (s, 1H), 9.13 (s, 2H), 7.41 (s, 2H), 7.14–7.08 (m, 4H), 7.04 (d, J = 7.6 Hz, 2H), 6.75 (m, 3H), 4.43 (d, J = 13.1 Hz, 2H), 4.29 (d, J = 13.8 Hz, 2H), 4.18 (dt, J = 14.3, 7.0 Hz, 2H),

3.54 (d, J = 3.7 Hz, 2H), 3.51 (d, J = 3.0 Hz, 2H), 2.30–2.18 (m, 2H), 0.93 (t, J = 6.9 Hz, 3H). ^{13}C NMR (CDCl$_3$, 100 MHz): δ 157.12, 153.92, 152.75, 143.34, 134.69, 133.88, 133.83, 132.95, 132.80, 132.69, 132.46, 130.99, 130.72, 130.29, 129.83, 129.48, 129.09, 128.73, 127.91, 127.69, 127.39, 126.82, 124.75, 120.56, 120.26, 119.84, 118.09, 117.38, 110.75, 80.54, 31.94, 31.45, 23.22, 23.22, 10.38. ESI-MS calcd for [M−H]$^-$ C$_{32}$H$_{28}$NO$_4$ 490.20, found 490.48.

Synthesis of 11: To a suspension of 0.500 g (1.02 mmol) of mono-propyl ether 10 and 1,8-bis(dimethylamino)naphthalene (proton sponge) (0.567 g, 2.65 mmol) in dry CH$_2$Cl$_2$ (20 mL) at 0 °C, trifluoromethanesulphonic anhydride (0.34 mL, 2.04 mmol) was added under nitrogen. After 2 h of stirring at room temperature, the organic layer was washed twice with 10% HCl and once with water, dried over anhydrous MgSO4, and evaporated. The residue was subjected to column chromatography purification (CH$_2$Cl$_2$/Hexane 3/10 v/v) giving product 11. Yield: 0.439 g (69%); Off-white solid; 1H NMR (CD$_2$Cl$_2$, 400 MHz): δ 7.33 (s, 2H), 7.23 (dd, J = 7.5, 1.3 Hz, 2H), 7.11 (dd, J = 7.6, 1.5 Hz, 2H), 6.96–6.84 (m, 3H), 6.81 (t, J = 7.5 Hz, 2H), 6.73 (s, 2H), 4.49 (d, J = 13.5 Hz, 2H), 4.18 (t, J = 6.5 Hz, 2H), 4.08 (d, J = 14.1 Hz, 2H), 3.61 (d, J = 14.1 Hz, 2H), 3.52 (d, J = 13.5 Hz, 2H), 2.38–2.12 (m, 2H), 1.34 (t, J = 7.4 Hz, 3H). 19F NMR: −74.21 (s). ESI-MS calcd for [M+Na]$^+$ C$_{33}$H$_{28}$F$_3$NNaO$_6$S 646.15, found 646.49.

Synthesis of carbazole-appended calix[4]arene 3c: To a mixture of Pd2dba3 (0.05 equiv.) and P(t-Bu)$_3$H$^+$BF$_4$$^-$ (0.2 equiv.) dissolved in 10 mL of dry DMF, CuI (2.5 equiv.), DBU (4 equiv.), carbazole alkyne (5 equiv.) and triflate 11 (0.25 mmol) were added and the mixture was heated at 85 °C in an oil bath for 12 h. The solvent was evaporated, and the resulting crude product was dissolved in CH$_2$Cl$_2$ and washed with brine several times. Drying the CH$_2$Cl$_2$ extract over anhydrous MgSO$_4$ followed by solvent removal under vacuum gave the crude product. The residue was subjected to column chromatography (CH$_2$Cl$_2$/Hexane 4/10 v/v) to obtain the pure compound. Yield: 0.053 g (30%); Off-white solid; ^{1}H NMR (CD$_2$Cl$_2$, 400 MHz): δ 8.53 (s, 1H), 8.20 (d, J = 7.7 Hz, 1H), 7.89 (d, J = 7.5 Hz, 1H), 7.54 (t, J = 7.0 Hz, 2H), 7.35–7.23 (m, 5H), 7.13 (d, J = 7.9 Hz, 2H), 7.04 (s, 2H), 6.97 (s, 3H), 6.85–6.65 (m, 3H), 4.89 (d, J = 13.4 Hz, 2H), 4.39 (t, J = 7.0 Hz, 2H), 4.29 (d, J = 13.5 Hz, 2H), 3.95 (t, J = 6.3 Hz, 2H), 3.77 (d, J = 13.5 Hz, 2H), 3.49 (d, J = 13.5 Hz, 2H), 1.98 (m, 2H), 1.85 (m, 2H), 1.04 (t, J = 7.3 Hz, 3H), 0.87 (t, J = 7.4 Hz, 3H). ^{13}C NMR (CD2Cl2, 100 MHz): δ 155.88, 153.60, 141.62, 141.08, 140.60, 135.31, 133.72, 133.48, 133.22, 130.78, 129.47, 129.42, 129.30, 128.84, 128.56, 128.24, 128.06, 127.93, 127.32, 126.63, 126.23, 126.13, 123.92, 122.47, 122.07, 120.61, 120.56, 119.79, 119.43, 118.52, 113.23, 109.24, 109.06, 99.82, 85.68, 80.50, 78.92, 46.66, 44.82, 36.38, 31.27, 23.39, 22.38, 11.56, 10.29. HRMS (ESI-TOF) m/z [M+H]$^+$ calcd for C$_{49}$H$_{43}$N$_2$O$_3$ 707.3274, found 707.3279.

Synthesis of 13: Calixarene 12 [24] (2.00 g, 2.87 mmol) was suspended in CH$_3$CN (40 mL), and after addition of 48% HBr (10 mL) the mixture was stirred at 60 °C for 12 h. The resulting suspension was diluted with CH$_2$Cl$_2$ and washed twice with water and once with brine. The organic layer dried over MgSO$_4$ and evaporated. The residue was washed several times with MeOH to remove benzyl alcohol giving the free phenol. The residue was recrystallized from CH$_3$Cl/MeOH to give pure calixarene 13. Yield: 1.04 g (70%); Pale yellow solid; ^{1}H NMR (CDCl$_3$, 400 MHz): δ 9.76 (d, J = 1.0 Hz, 1H), 9.58 (d, J = 1.0 Hz, 1H), 9.38 (d, J = 1.1 Hz, 1H), 7.21 (d, J = 1.0 Hz, 2H), 7.16–7.03 (m, 5H), 7.01–6.92 (m, 2H), 6.80–6.70 (m, 2H), 4.48 (d, J = 12.8 Hz, 1H), 4.36–4.21 (m, 3H), 4.20–4.05 (m, 2H), 3.63–3.25 (m, 4H), 2.31–2.13 (m, 2H), 1.32 (td, J = 7.3, 1.3 Hz, 3H). ^{13}C NMR (CDCl$_3$, 100 MHz): δ 150.56, 150.51, 150.43, 149.38, 134.09, 134.06, 131.20, 131.04, 130.88, 130.69, 129.73, 129.16, 128.90, 128.72, 128.43, 128.37, 128.02, 126.36, 125.59, 125.49, 122.17, 121.47, 119.16, 112.28, 79.24, 31.94, 31.89, 31.86, 3.31, 31.04, 23.38, 10.77. ESI-MS calcd for [M-H]$^-$ C$_{31}$H$_{28}$BrO$_4$ 543.12, found 543.50.

Synthesis of 14: To a suspension of 1.00 g (1.93 mmol) of mono-propyl ether 13 and 1,8-bis(dimethylamino)naphthalene (proton sponge) (1.08 g, 5.02 mmol) in dry CH$_2$Cl$_2$ (40 mL) at 0 °C, trifluoromethanesulphonic anhydride (0.65 mL, 3.86 mmol) was added under nitrogen. After 2 h of stirring at room temperature, the organic layer was washed twice with 10% HCl and once with water, dried over anhydrous MgSO4, and evaporated.

The residue was subjected to column chromatography purification (CH_2Cl_2/Hexane 3/10 *v/v*) giving product 14.

Yield: 0.938 g (75%); White solid; ^{1}H NMR ($CDCl_3$, 400 MHz): δ 7.43 (s, 1H), 7.30 (s, 2H), 7.27–7.10 (m, 4H), 6.99 (dd, J = 13.1, 7.5 Hz, 2H), 6.91–6.73 (m, 4H), 4.48 (dd, J = 12.6, 9.5 Hz, 2H), 4.17 (t, J = 6.2 Hz, 2H), 3.99 (t, J = 13.0 Hz, 2H), 3.69–3.37 (m, 4H), 2.29–2.06 (m, 2H), 0.92 (t, J = 6.0 Hz, 3H). ^{13}C NMR ($CDCl_3$, 100 MHz): δ 152.90, 152.18, 149.91, 134.13, 133.11, 132.69, 131.75, 131.28, 131.13, 130.92, 130.32, 130.00, 129.95, 129.63, 128.96, 128.72, 127.62, 127.18, 125.52, 119.72, 111.09, 80.34, 31.93, 31.72, 31.49, 23.18, 22.78, 14.24, 10.53. 19F NMR: -74.34 (s). ESI-MS calcd for [M+Na]$^+$ $C_{32}H_{28}BrF_3O_6SNa$ 699.06, found 699.52.

Synthesis of 15: To a mixture of Pd_2dba_3 (0.05 equiv.) and P(t-Bu)$_3$H$^+$ BF$_4^-$ (0.2 equiv.) dissolved in 10 mL of dry DMF, CuI (2.5 equiv.), DBU (4 equiv.), carbazole alkyne (5 equiv.) and triflate 14 (0.25 mmol) were added and the mixture was heated at 85 °C in an oil bath for 12 h. The solvent was evaporated, and the resulting crude product was dissolved in CH_2Cl_2 and washed with brine several times. Drying the CH_2Cl_2 extract over anhydrous $MgSO_4$ followed by solvent removal under vacuum gave the crude product. The residue was subjected to column chromatography (CH_2Cl_2/Hexane 4/10 *v/v*) to obtain the pure compound. Yield: 0.086 g (45%); Yellow solid; ^{1}H NMR ($CDCl_3$, 400 MHz): δ 8.47 (s, 1H), 8.18 (d, J = 7.8 Hz, 1H), 7.85 (d, J = 8.4 Hz, 1H), 7.64–7.42 (m, 2H), 7.34 (t, J = 9.3 Hz, 1H), 7.26–7.08 (m, 6H), 7.07–6.92 (m, 3H), 6.80 (m, 5H), 4.90 (t, J = 14.0 Hz, 2H), 4.35 (dd, J = 13.6, 6.6 Hz, 2H), 4.26–3.98 (m, 2H), 3.93 (t, J = 6.4 Hz, 2H), 3.66 (dd, J = 23.6, 11.9 Hz, 2H), 3.43 (dd, J = 25.8, 12.1 Hz, 2H), 1.99 (dd, J = 14.5, 7.2 Hz, 2H), 1.82 (dd, J = 14.0, 7.0 Hz, 2H), 1.03 (t, J = 7.5 Hz, 3H), 0.86 (t, J = 7.3 Hz, 3H). ^{13}C NMR ($CDCl_3$, 100 MHz): δ 153.03, 151.69, 142.00, 141.07, 140.80, 140.46, 133.37, 132.40, 131.51, 131.47, 131.04, 130.97, 130.74, 130.35, 130.23, 129.86, 129.78, 129.68, 129.51, 129.40, 129.29, 129.13, 128.49, 128.28, 128.12, 127.84, 127.60, 127.46, 127.41, 127.25, 126.48, 126.15, 125.84, 124.09, 120.82, 120.50, 119.45, 114.08, 110.94, 109.11, 108.83, 99.00, 86.22, 80.37, 78.66, 44.92, 37.58, 37.35, 36.64, 36.55, 31.95, 31.85, 23.42, 23.36, 22.50, 11.94, 10.64. ESI-MS calcd for [M-H]$^-$ $C_{48}H_{41}BrNO_3$ 758.23, found 758.50.

Synthesis of carbazole-appended calix[4]arene 3d: An oven-dried Schlenk tube was charged with $Pd_2(dba)_3$ (0.05 equiv.), JohnPhos (0.1 equiv.), calix halide 15 (0.08 mmol), amine (0.16 mmol) and toluene (2 mL). The reaction was stirred for few minutes and then LiHMDS (0.9–1.1 M in Hexanes) (0.18 mL) was added via syringe. The reaction vessel was sealed and heated at 80 °C with stirring for 48 h. The reaction mixture was then allowed to cool to room temperature, adsorbed on silica, and purified by column chromatography with EtOAc/hexane (2/8 *v/v*). Yield: 0.013 g (20%); Off-white solid; ^{1}H NMR (CD_2Cl_2, 400 MHz): δ 8.33 (s, 1H), 8.16 (d, J = 7.8 Hz, 1H), 7.68 (d, J = 8.4 Hz, 1H), 7.58–7.45 (m, 3H), 7.42 (d, J = 6.0 Hz, 2H), 7.30 (t, J = 7.2 Hz, 1H), 7.23–7.13 (m, 2H), 7.10 (s, 1H), 7.02–6.76 (m, 4H), 6.66 (s, 1H), 6.56–6.37 (m, 3H), 4.53 (dt, J = 12.1, 5.9 Hz, 2H), 4.35 (t, J = 7.1 Hz, 2H), 4.15 (d, J = 14.1 Hz, 2H), 4.06 (t, J = 6.0 Hz, 2H), 3.58–3.43 (m, 4H), 3.39–2.83 (m, 4H), 2.09–1.93 (m, 4H), 1.81 (s, 4H), 1.72 (s, 2H), 1.26 (t, J = 7.2 Hz, 3H), 1.03 (t, J = 7.3 Hz, 3H). ^{13}C NMR (CD_2Cl_2, 100 MHz): δ 153.76, 153.37, 151.62, 149.13, 141.05, 140.08, 138.63, 138.11, 133.07, 132.51, 131.60, 131.09, 130.75, 130.60, 129.40, 129.27, 129.18, 129.06, 128.41, 128.31, 127.67, 127.43, 126.15, 125.56, 125.44, 123.56, 123.48, 122.87, 122.41, 120.47, 119.29, 119.03, 114.01, 113.83, 109.15, 109.04, 88.59, 88.02, 78.62, 44.82, 33.93, 33.75, 31.44, 31.24, 27.71, 25.09, 23.35, 22.38, 13.81, 11.57, 10.43. HRMS (ESI-TOF) m/z [M]$^+$ calcd for $C_{53}H_{53}N_2O_3$ 765.4056, found 765.4057.

4. Conclusions

In summary, we presented the first studies of the electronic effects on the host–guest complexation in fluorescent calixarene scaffolds. We found that the introduction of an electron-donating substituent in the aromatic ring opposing the fluorophore-substituted ring enhances the complexation of the cationic N-methyl pyridinium guest, while an electron-withdrawing substituent in the same position decreases this complexation. In contrast, an electronic donor at the ring adjacent to the fluorophore-substituted one does not provide stronger binding of the cationic guest. Thus, our results provide strong evidence for the planar cationic guest undergoing π interactions with only one pair of the calixarene aromatic rings which is not involved in the hydrogen bonding at the lower rim. Although more electron rich, this hydrogen bonding between the unsubstituted phenolic rings is likely too strong to make them available for the π interactions.

Supplementary Materials: Synthesis and characterization of all new compounds, UV-vis, fluorescence and NMR spectra, fluorescence and NMR complexation studies. This information can be downloaded at: https://www.mdpi.com/article/10.3390/molecules27175689/s1. Reference [28] has been cited the Supplementary Materials.

Author Contributions: Conceptualization, writing—review and editing, A.V.; formal analysis and investigation, V.R. All authors have read and agreed to the published version of the manuscript.

Funding: We thank the Israel Science Foundation (ISF, grant 1328/20) and BIKURA (administered by ISF, grant 1928/19) for supporting this work.

Institutional Review Board Statement: Not applicable.

Informed Consent Statement: Not applicable.

Data Availability Statement: All experimental data are provided in the Supplementary information.

Conflicts of Interest: The authors declare no conflict of interest.

Sample Availability: Samples of the compounds are not available from the authors.

References

1. Abraham, W. Inclusion of Organic Cations by Calix[n]arenes. *J. Incl. Phenom. Macrocyc. Chem.* **2002**, *43*, 159–174. [CrossRef]
2. Dalla Cort, A.; Mandolini, L. *Calixarenes in Action*; Mandolini, L., Ungaro, R., Eds.; Imperial College Press: London, UK, 2000.
3. Schalley, C.A.; Castellano, R.K.; Brody, M.S.; Rudkevich, D.M.; Siuzdak, G.; Rebek, J. Investigating Molecular Recognition by Mass Spectrometry: Characterization of Calixarene-Based Self-Assembling Capsule Hosts with Charged Guests. *J. Am. Chem. Soc.* **1999**, *121*, 4568–4579. [CrossRef]
4. Lhotak, P.; Shinkai, S. Cation–π Interactions in Calix[n]arene and Related Systems. *J. Phys. Org. Chem.* **1997**, *10*, 273–285. [CrossRef]
5. Shinkai, S.; Araki, K.; Manabe, O. NMR Determination of Association Constants for Calixarene Complexes. Evidence for the Formation of a 1:2 Complex with Calix[8]arene. *J. Am. Chem. Soc.* **1988**, *110*, 7214–7215. [CrossRef]
6. Rebek, J., Jr. Host–guest Chemistry of Calixarene Capsules. *Chem. Commun.* **2000**, *8*, 637–643. [CrossRef]
7. Araki, K.; Shimizu, H.; Shinkai, S. Cation-π Interactions in Calix[4]arene-based Host Molecules. What Kind of Cavity-shape Is Favored for the Cation-binding? *Chem. Lett.* **1993**, *22*, 205–208. [CrossRef]
8. Araki, K.; Hisaichi, K.; Kanai, T.; Shinkai, S. Synthesis of an Upper-rim-connected Biscalix[4]arene and Its Improved Inclusion Ability Based on the Cooperative Action. *Chem. Lett.* **1995**, *24*, 569–570. [CrossRef]
9. Arduini, A.; Pochini, A.; Secchi, A. Rigid Calix[4]arene as a Building Block for the Synthesis of New Quaternary Ammonium Cation Receptors. *Eur. J. Org. Chem.* **2000**, *2000*, 2325–2334. [CrossRef]
10. Pescatori, L.; Arduini, A.; Pochini, A.; Secchi, A.; Massera, C.; Ugozzoli, F. Monotopic and heteroditopic calix[4]arene receptors as hosts for pyridinium and viologen ion pairs: A solution and solid-state study. *Org. Biomol. Chem.* **2009**, *7*, 3698–3708. [CrossRef]
11. Ludwig, R.; Dzung, N.T.K. Calixarene-Based Molecules for Cation Recognition. *Sensors* **2002**, *2*, 397–416. [CrossRef]
12. Zgadzaj, M.O.; Wendel, V.; Fehlinger, M.; Ziemer, B.; Abraham, W. Inclusion of Organic Cations by Calix[4]arenes Bearing Cyclohepta-2,4,6-trienyl Substituents. *Eur. J. Org. Chem.* **2001**, *2001*, 1549–1561. [CrossRef]
13. Dyker, G.; Mastalerz, M.; Müller, I.M. Electron-Rich Cavitands via Fourfold Sonogashira Cross-Coupling Reaction of Calix[4]arenes and Bromopyridines—Effect of the Nitrogen Position on Complexation Abilities. *Eur. J. Org. Chem.* **2005**, *2005*, 3801–3812. [CrossRef]
14. Puchnin, K.; Cheshkov, D.; Zaikin, P.; Vatsouro, I.; Kovalev, V. Tuning conformations of calix[4]tubes by weak intramolecular interactions. *New J. Chem.* **2013**, *37*, 416–424. [CrossRef]
15. Ikeda, A.; Shinkai, S. Novel Cavity Design Using Calix[*n*]arene Skeletons: Toward Molecular Recognition and Metal Binding. *Chem. Rev.* **1997**, *97*, 1713–1734. [CrossRef]
16. de Namor, A.F.D.; Cleverley, R.M.; Zapata-Ormachea, M.L. Thermodynamics of Calixarene Chemistry. *Chem. Rev.* **1998**, *98*, 2495–2526. [CrossRef]
17. Arduini, A.; Fabbi, M.; Mantovani, M.; Mirone, L.; Pochini, A.; Secchi, A.; Ungaro, R. Calix[4]arenes Blocked in a Rigid Cone Conformation by Selective Functionalization at the Lower Rim. *J. Org. Chem.* **1995**, *60*, 1454–1457. [CrossRef]
18. Soi, A.; Bauer, W.; Mauser, H.; Moll, C.; Hampel, F.; Hirsch, A. Investigations on the dynamic properties of 25,26,27,28-tetraalkoxycalix[4]arenes: Para-substituent- and solvent-dependent properties of paco conformers and determination of thermodynamic parameters of the pinched cone/pinched cone conversion. *J. Chem. Soc. Perkin Trans.* **1998**, *2*, 1471–1478. [CrossRef]

19. Al-Saraierh, H.; Miller, D.O.; Georghiou, P.E. For the first report on the Sonogashira coupling at the lower rim and relevant X-ray structures see: Narrow-Rim Functionalization of Calix[4]arenes via Sonogashira Coupling Reactions. *J. Org. Chem.* **2005**, *70*, 8273–8280. [CrossRef]
20. Molad, A.; Goldberg, I.; Vigalok, A. Tubular Conjugated Polymer for Chemosensory Applications. *J. Am. Chem. Soc.* **2012**, *134*, 7290–7292. [CrossRef]
21. Neri, P.; Bottino, A.; Cunsolo, F.; Piattelli, M.; Gavuzzo, E. For the original synthesis of the 5,5′-Bicalixarene scaffold see: 5,5′-Bicalix[4]arene: The Bridgeless Prototype of Double Calix[4]arenes of the "Head-to-Head" Type. *Angew. Chem. Int. Ed.* **1998**, *37*, 166–169. [CrossRef]
22. Ahuja, B.B.; Vigalok, A. Fluorescent Calixarene Scaffolds for NO Detection in Protic Media. *Angew. Chem. Int. Ed.* **2019**, *58*, 2774–2778. [CrossRef] [PubMed]
23. Baheti, A.; Dobrovetsky, R.; Vigalok, A. Fluorophore-Appendant 5,5′-Bicalixarene Scaffolds for Host–Guest Sensing of Nitric Oxide. *Org. Lett.* **2020**, *22*, 9706–9711. [CrossRef] [PubMed]
24. Vézina, M.; Gagnon, J.; Villeneuve, K.; Drouin, M.; Harvey, P.D. (η^5-Pentamethylcyclopentadienyl)rhodium Complexes of Upper-Rim Monophosphinated Calix[4]arene. *Organometallics* **2001**, *20*, 273–281. [CrossRef]
25. Shimizu, S.; Moriyama, A.; Kito, A.K.; Sasaki, Y. Selective Synthesis and Isolation of All Possible Conformational Isomers of Proximally Para-Disubstituted Calix[4]arene. *J. Org. Chem.* **2003**, *68*, 2187–2194. [CrossRef]
26. Shirakawa, S.; Kimura, T.; Murata, S.; Shimizu, S. Synthesis and Resolution of a Multifunctional Inherently Chiral Calix[4]arene with an ABCD Substitution Pattern at the Wide Rim: The Effect of a Multifunctional Structure in the Organocatalyst on Enantioselectivity in Asymmetric Reactions. *J. Org. Chem.* **2009**, *74*, 1288–1296. [CrossRef]
27. Harris, M.C.; Huang, X.; Buchwald, S.L. Improved Functional Group Compatibility in the Palladium-Catalyzed Synthesis of Aryl Amines. *Org. Lett.* **2002**, *4*, 2885–2888. [CrossRef]
28. Fielding, L. Determination of Association Constants (K_{ass}) from Solution NMR Data. *Tetrahedron* **2000**, *34*, 6151–6170. [CrossRef]

Article

The Formation of Inherently Chiral Calix[4]quinolines by Doebner–Miller Reaction of Aldehydes and Aminocalixarenes

Martin Tlustý [1], Václav Eigner [2], Hana Dvořáková [3] and Pavel Lhoták [1,*]

1   Department of Organic Chemistry, University of Chemistry and Technology, Prague (UCTP), Technická 5, 166 28 Prague, Czech Republic
2   Department of Solid State Chemistry, UCTP, Technická 5, 166 28 Prague, Czech Republic
3   Laboratory of NMR Spectroscopy, UCTP, Technická 5, 166 28 Prague, Czech Republic
*   Correspondence: lhotakp@vscht.cz; Tel.: +420-440225055

Abstract: The formation of inherently chiral calix[4]arenes by the intramolecular cyclization approach suffers from a limited number of suitable substrates for these reactions. Here, we report an easy way to prepare one class of such compounds: calixquinolines, which can be obtained by the reaction of aldehydes with easily accessible aminocalix[4]arenes in acidic conditions (Doebner–Miller reaction). The synthetic procedure represents a very straightforward approach to the inherently chiral macrocyclic systems. The complexation studies revealed the ability of these compounds to complex quaternary ammonium salts with different stoichiometries depending on the guest molecules. At the same time, the ability of enantioselective complexation of chiral *N*-methylammonium salts was demonstrated.

Keywords: calixarene; inherent chirality; complexation; mercuration; *meta*-substitution; quinoline formation; chiral recognition

Citation: Tlustý, M.; Eigner, V.; Dvořáková, H.; Lhoták, P. The Formation of Inherently Chiral Calix[4]quinolines by Doebner–Miller Reaction of Aldehydes and Aminocalixarenes. *Molecules* **2022**, *27*, 8545. https://doi.org/10.3390/molecules27238545

Academic Editor: Paula M. Marcos

Received: 31 October 2022
Accepted: 1 December 2022
Published: 4 December 2022

Publisher's Note: MDPI stays neutral with regard to jurisdictional claims in published maps and institutional affiliations.

1. Introduction

Calixarenes [1–4] are macrocyclic compounds which are widely used in supramolecular chemistry. The reason for their popularity can be found in the combination of several factors: easy multi-gram preparation, variable size and tuneable shape of the cavity, simple derivatization, etc. Depending on the substitution, calixarenes exhibit good complexation properties towards cations, anions or neutral compounds [5–9]. However, unlike some other macrocycles (e.g., cyclodextrins, pillararenes), calixarenes themselves are achiral molecules, which makes them useless as chiral receptors without further derivatization. One possible solution to chirality issues is to convert calixarenes into so-called inherently chiral systems [10–12], where the combination of nonplanar molecules with suitable substitution patterns can lead to chirality without the introduction of stereogenic units. For example, calix[4]arenes in the *cone* conformation with WXYZ (Figure 1a) or XY (Figure 1b) substitution patterns (upper rim or lower rim) exhibit such chirality because the presence of different substituents gives the system a particular sense of rotation (Figure 1a). Similarly, the *meta* substitution [13] of calix[4]arenes yields chiral compounds (Figure 1c) and, as such, represents the most straightforward approach to such systems. However, an obvious drawback of this approach is the lack of corresponding derivatization techniques enabling selective substitution of the *meta* position. In fact, the electrophilic aromatic mercuration is so far the only procedure known to provide *meta* substituted products from an unsubstituted calix[4]arenes [14].

Intramolecular cyclization represents a different way for the synthesis of inherently chiral *meta* substituted calixarenes. This approach consists of two steps: (i) the introduction of a suitable functional group to the *para* position of calixarene; and (ii) intramolecular cyclization, leading to calixarenes with fused rings. As an example, we can mention the preparation of the naphthalene moiety (A) starting from an aldehyde [15], phenanthrene moiety (B) by the photocyclization of stilbene [16], the formation of calixarene-fused phosphols [17] (C) and the preparation of calixquinazolines (D) (Figure 2) [18]. However, most of these approaches suffer from a complicated synthesis of starting compounds or a low-yielding cyclization step. Miao et al. [19] also synthesized calixquinolines (E); however, this approach required the utilization of bifunctional reagents (crotonaldehyde or ethyl acetoacetate).

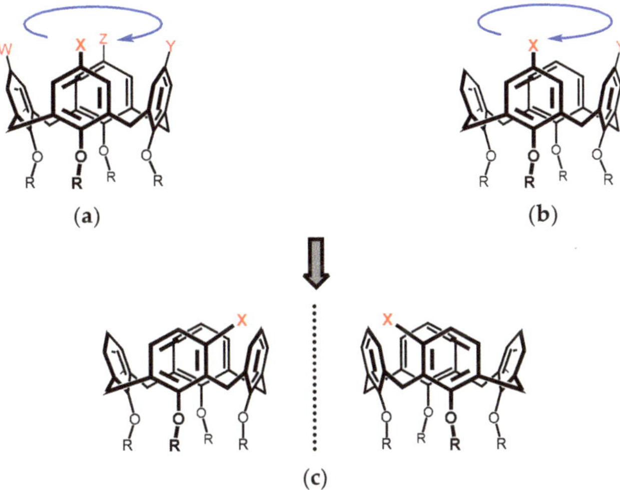

Figure 1. Examples of inherently chiral calix[4]arenes: (**a**) WXYZ pattern; (**b**) XY pattern; (**c**) *meta* substitution.

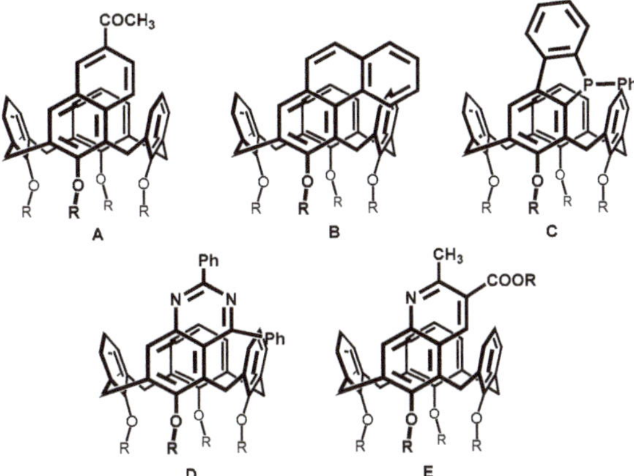

Figure 2. Examples of calix[4]arenes with fused rings.

In this paper, we report on the synthesis of calixquinolines from easily accessible aminocalix[4]arenes derivatives (both *meta-* and *para-*) in the *cone* conformation by the tandem reaction with simple aldehydes (acetaldehyde, bromoacetaldehyde). Although the reaction conditions were initially designed for the Pictet–Spengler condensation [20,21] of *meta*-aminocalixarene with aldehydes, the unexpected formation of calixquinolines (Doebner–Miller reaction) turned out to be general and worked with *para*-aminocalixarenes as well. This synthetic procedure represents a very straightforward approach to calixquinoline derivatives representing the inherently chiral macrocyclic systems.

2. Results and Discussion

Our original intention was to prepare amine-bridged macrocycles **5** starting from the *meta*-aminocalix[4]arenes (Scheme 1). These compounds would represent a reduced form of our recently reported imine-bridged calixarenes [22,23]. Using the known procedure, 4-aminocalix[4]arenes **2** was prepared in three steps from the starting calixarene **1** (Scheme 1). Thus, the initial mercuration of **1** with one equivalent of $Hg(TFA)_2$ gave the corresponding *meta* HgCl derivative in 65% yield [14]. Subsequent reaction with isoamyl nitrite/HCl gave nitroso compound (91%) [24], which was finally reduced by N_2H_4/Ni(R) to the desired amine **2** (89%) [22].

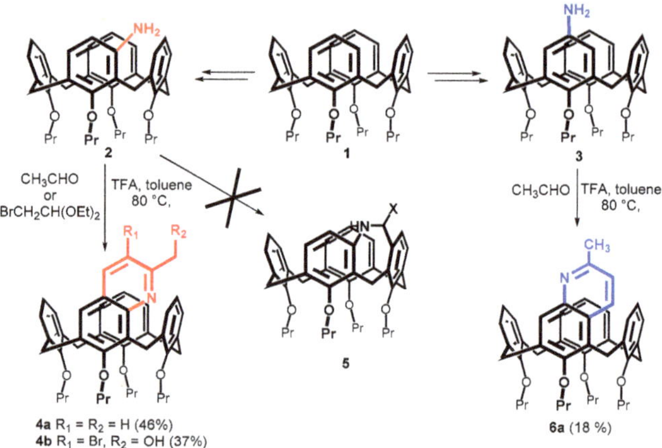

Scheme 1. Synthesis of calix[4]quinolines from *meta*- and *para*-aminosubstituted calix[4]arenes.

Surprisingly, the Pictet–Spengler reaction of calixarene **2** with aliphatic aldehydes did not provide the expected amine-bridged compounds **5**. Conversely, the reaction of amine **2** with acetaldehyde in the presence of TFA in toluene at 80 °C gave the quinoline derivative **4a** in 46% yield (Scheme 1). Similarly, the same reaction conditions and bromoacetaldehyde diethyl acetal gave compound **4b** in 37% yield. The unexpected hydroxy group was probably introduced into the molecule by the substitution of the bromine atom during the workup. The formation of these compounds can be explained by a Doebner–Miller reaction [25,26] consisting of the initial aldol reaction, followed by the conjugate addition and the final ring closure. Indeed, benzaldehyde, which does not work as a substrate in aldol reactions, provided a complex mixture of products. The same holds for the ketones acetone and acetophenone. These results indicated that only aliphatic aldehydes are suitable for this type of reaction.

To test the general applicability within the calixarene series, the *para*-amino-substituted derivative **3** was prepared by the nitration of starting **1** with 100% HNO_3 in the presence of glacial acetic acid in dichloromethane [27] and the subsequent reduction of nitro intermediate with $SnCl_2$ [28]. The reaction with acetaldehyde carried out under identical conditions as for **2** (TFA in toluene at 80 °C) provided quinoline derivative **6a** in 18% yield (Scheme 1).

The structure of compound **4a** was confirmed by the combination of NMR and HRMS techniques. The ^1H NMR spectrum (400 MHz, $CDCl_3$) showed the presence of two sets of four doublets for the methylene bridge protons at 5.03, 4.66, 4.52, 4.51 and 4.35, 3.37 and 3.19 (2×) with typical geminal coupling constants (~13.5 Hz) consistent with the C_1-symmetrical calix[4]arenes. The spectrum also contained a singlet at 2.76 ppm, revealing the presence of a methyl group. Moreover, the presence of a significantly downfield-shifted doublet at 8.00 ppm suggested the presence of a strong electron-withdrawing group (nitrogen) within the aromatic structure. The HRMS ESI+ analysis showed signals at m/z = 658.3892 and 680.3705 corresponding to [M+H]+ (658.3891) and [M+Na]+ (680.3710) predicted for **4a**.

As the above synthetic protocol represents a very straightforward approach to inherently chiral calixarene derivatives, we decided to prepare a library of diaminocalixarenes, which are synthetically available (Scheme 2).

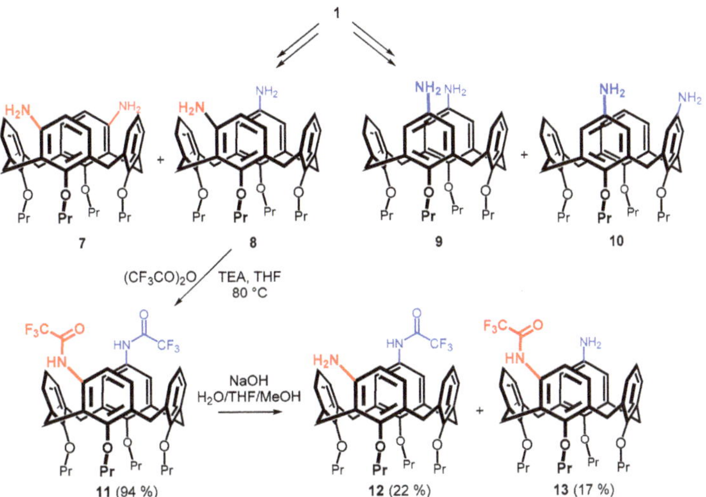

Scheme 2. Synthesis of *meta-* and *para*-aminosubstituted calix[4]arenes.

The starting tetrapropoxycalix[4]arenes **1** was transformed into diamines **7** and **8** using the procedure described by our group. Briefly, a reaction with two equivalents of Hg(TFA)$_2$ in chloroform provided a mixture of two distally dimercurated calixarenes: *meta,meta* and *meta,para* isomers [29]. The mixture was reacted with isoamyl nitrite in the presence of HCl and AcOH in chloroform to yield the corresponding nitroso derivatives, which are easily separated by column chromatography on silica gel. The resulting amines **7** and **8** were obtained by reduction with hydrazine in the presence of Raney-nickel in refluxing ethanol [23]. To synthesize the *para*-diamino derivatives **9** and **10**, calix[4]arene **1** was nitrated with 100% HNO$_3$ in glacial acetic acid and dichloromethane as a solvent [27]. The corresponding distal and proximal dinitro derivatives were separated by column chromatography and finally reduced by SnCl$_2$ in refluxing toluene [28].

Compound **8** represents a rather unusual structural motif as it contains both *para-* and *meta*-amino groups within the molecule. In order to study possible differences in the reactivity of the two regioisomeric groups, the substance was reacted with an excess of (CF$_3$CO)$_2$O in the presence of TEA in THF to form diamide **11** in 94% yield (Scheme 2). A careful hydrolysis of this compound finally gave monoamines **12** (*meta*) and **13** (*para*) in 22 and 17% yield, respectively, after a column chromatography on silica gel [23].

Not all of the amino derivatives shown in Scheme 2 proved useful for the preparation of calixquinolines. Thus, the reaction of acetaldehyde with calixarene **7** (*meta, meta* substitution) gave quinoline **14** in 37% yield (Scheme 3). Unlike the previous example, the cyclization of amine **8** provided a mixture of the two regioisomers/diastereoisomers **15a** and **15b** in low yield (14%). Unfortunately, the mixture was not separable using conventional separation techniques (column chromatography, preparative TLC, flash chromatography). On the other hand, the unexpected by-product **15c** possessing ethylamino group in the *para*-position was also isolated in 10% yield.

Although cyclization of diamines **9** and **10** should provide only 2 or 3 regioisomers, respectively, in both cases the reaction with acetaldehyde turned out to be much more complicated, and we were unable to isolate any expected quinoline product from complex reaction mixtures. In contrast, *meta*-amine **12** smoothly provided quinoline **16** in 47% yield (Scheme 3). The cyclization of its *para-* congener **13** resulted in the formation of two regioisomers **17a** and **17b**, which were isolated by column chromatography on silica gel in 22% and 17% yields, respectively. The unambiguous proof of the structure of **16** was obtained from single crystal X-ray analysis. The monocrystals of **16** were obtained from dichloromethane/MeOH mixture as a methanol solvate (1:1) in the triclinic system, *P-1* space group. As shown in Figure 3a,b the macrocycle adopts the *pinched cone* conformation, which is common for solid-state structures of the *cone* isomers. Defining the main plane of the calixarene by the four bridging carbon atoms (C2, C8, C14, C20), the aromatic moiety bearing trifluoroacetamide is tilted out of the cavity with the interplanar angle $\Phi = 131.40°$. The opposite aromatic subunit with quinoline moiety exhibits similar geometrical parameters ($\Phi = 127.36°$). The remaining two phenolic subunits cross the main plane at almost right angles ($\Phi = 79.48°$ and $84.75°$) and are directed slightly into the cavity. This arrangement with large substituents on the subunits facing out of the cavity is apparently the result of steric hindrance minimization (Figure 3b).

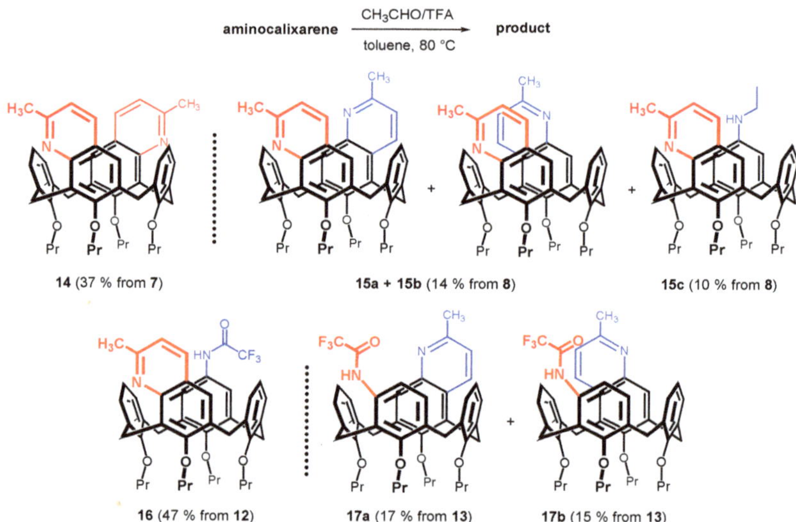

Scheme 3. Synthesis of calix[4]quinolines (the number of starting aminocalixarene in brackets); reaction conditions: acetaldehyde, TFA, toluene, 80 °C.

Interesting supramolecular interactions were found within the crystal packing of **16**. As shown in Figure 3c, the opposite enantiomers of **16** are held together via hydrogen bond interactions between the carbonyl oxygen and NH groups from amidic functions (NH···O = C distance = 2.128 Å). This bonding is further strengthened by the concomitant HB interaction between the carbonyl and *ortho* hydrogen of aromatic subunit (CH···O = C = 2.522 Å). Interestingly, methanol is included in the form of dimer with OH···O distance of 1.817 Å, indicating a strong HB between methanol molecules (Figure 3c). This dimer is held at both ends by HBs from quinoline nitrogen (N···HO = 2.129 Å) and from amidic NH moiety (NH···O = 1.927 Å).

Common lower rim peralkylated calix[4]arenes immobilized in the *cone* conformation are known to exhibit so called *pinched cone—pinched cone* interconversion in solution, where two border conformations mutually equilibrate (Figure 4). The above X-ray analysis revealed that the introduction of heterocyclic moiety into the upper rim of calix[4]arenes leads to a single *pinched cone* conformer in the solid state with substituted aromatic subunits pointing out of the cavity. This means that only one of the two theoretically possible *pinched cone* conformations is preferred, obviously as a consequence of the steric hindrance minimization within the upper rim.

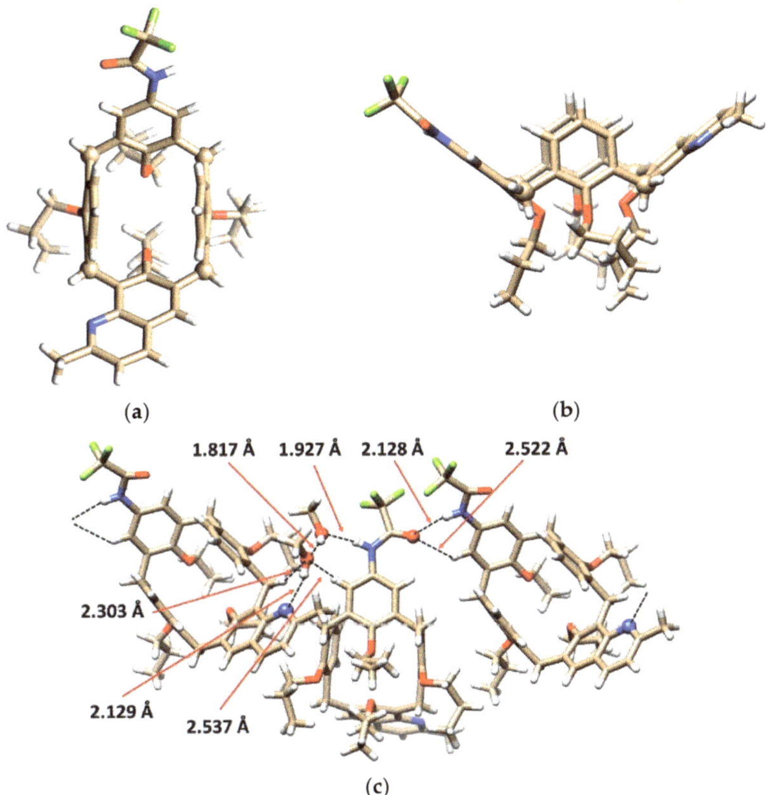

Figure 3. X-ray structure of **16**: (**a**) top view showing *pinched cone* conformation (bridging C atoms shown as balls for better clarity); (**b**) side view of the same; (**c**) the binding motif in crystal packing of enantiomers; methanol solvate was removed for better clarity.

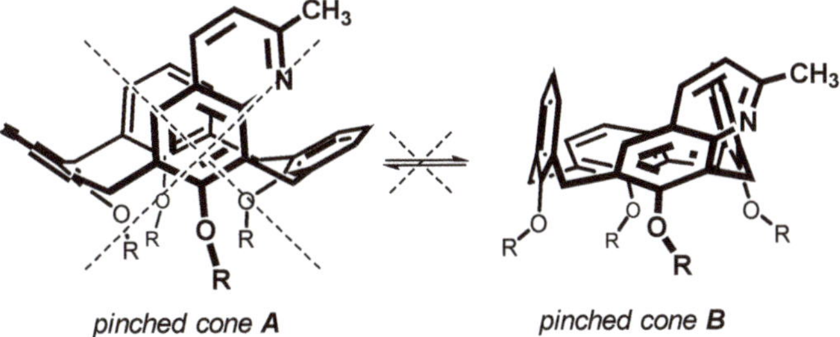

Figure 4. *Pinched cone—pinched cone* interconversion in calix[4]arenes (shown for compound **4a**).

To show the general behaviour of our products in solution, compounds **4a** and **6a** representing the *meta-* and *para-*amino substituted calixarene systems were selected for dynamic ^{1}H NMR study. As shown in ESI (Figures S45–S50), both compounds **4a** and **6a** did not show any changes within the whole temperature range studied (298-173 K, CD$_2$Cl$_2$, 500 MHz). This is strong evidence that both compounds also exist in solution in only one thermodynamically preferred *pinched cone B* conformation as a direct consequence of the attachment of the heterocyclic moiety.

As all quinoline derivatives described above represent the inherently chiral systems, we have carried out a study of their possible resolution on a chiral column. On an analytical scale, the separation of racemic **4a** into enantiomers was feasible with Chiralpak IA (250 × 4.6 ID, 5 µm) column, using heptane/ethyl acetate (95/5, *v/v*) as a mobile phase. However, these conditions turned out to be unsuitable for the separation in a preparative scale. On the other hand, the preparative resolution of compound **6a** was successfully carried out using a polysaccharide column ChiralArt Amylose-SA (250 × 20 mm ID, 5 µm) with cyclohexane/DCM: 92/8 *v/v* as a mobile phase. The resulting individual enantiomers **6a_1** and **6a_2** were used for titration study.

Calixquinolines represent systems with enlarged aromatic cavities potentially capable of interacting with molecules bearing an acidic CH_3 group (C-H$\cdots\pi$ interactions). The ^{1}H NMR titration experiments revealed the possible use of newly prepared compounds as receptors for ammonium salt complexation (Figure 5). The complexation constant was determined by analyzing the complexation-induced shifts (CIS) of host signals (hydrogen in position 3- of quinolinium moiety) using a nonlinear curve-fitting procedure (program BindFit) [30]. The titration ($C_2D_2Cl_4$) of **4a** with *N*-methylquinolinium iodide (NMQI) and *N*-methylisoquinolinium iodide (NMII) revealed the formation of 1:1 complexes with complexation constants 12.2 and 16.6 M^{-1}, respectively (Standard deviations for NMR titrations are generally estimated to be around 10%). Surprisingly, the titration by a structurally similar *N*-methylpyridinium iodide (NMPI) showed the formation of complexes with 2:1 stoichiometry and complexation constants K_{11} = 13.0 M^{-1} and K_{21} = 2798 M^{-1}.

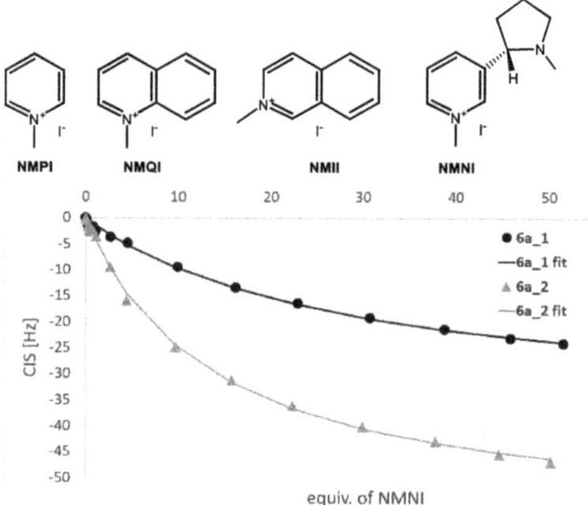

Figure 5. ^{1}H NMR titration curve of **6a_1** and **6a_2** with NMNI ($C_2D_2Cl_4$, 400 MHz, 298 K), hydrogen in position 3- of calixarene quinoline moiety observed.

Based on these results, we attempted the enantioselective recognition of (*S*)-*N*-methylnicotinium iodide (NMNI) with separated enantiomers **6a_1** and **6a_2** (Figure 4). The resulting complexation constants 37.3 M^{-1} for **6a_1** and 72.1 M^{-1} for **6a_2** showed the fairly good ability of our substances in enantioselective recognition of chiral guest molecules.

The presence of a trifluoroacetamide motif in several of our compounds led us to the idea of testing the anion-binding capacity as well. Indeed, compound **16** was shown to complex tetrabutylammonium acetate and benzoate in $CDCl_3$ solution with corresponding constants of $K_{(Ac)}$ = 23.6 M^{-1} and $K_{(Bz)}$ = 29.9 M^{-1}.

3. Materials and Methods

3.1. General Experimental Procedures

All chemicals were purchased from commercial sources and used without further purification. Solvents were dried and distilled using conventional methods. Melting points were measured on Heiztisch Mikroskop—Polytherm A (Wagner & Munz, Germany). NMR spectra were performed on Agilent 400-MR DDR2 (^{1}H: 400 MHz, ^{13}C: 100 MHz). Deuterated solvents used are indicated in each case. Chemical shifts (δ) are expressed in ppm and refer to the residual peak of the solvent or TMS as an internal standard; coupling constants (J) are in Hz. The mass analyses were performed using the ESI technique on a Q–TOF (Micromass) spectrometer. Elemental analyses were carried out on Perkin–Elmer 240, Elementar vario EL (Elementar, Germany) or Mitsubishi TOX–100 instruments. All samples were dried in the desiccator over P_2O_5 under vacuum (1 Torr) at 80 °C for 8 h. The IR spectra were measured on an FT–IR spectrometer Nicolet 740 or Bruker IFS66 spectrometers equipped with a heated Golden Gate Diamante ATR–Unit (SPECAC) in KBr. A total of 100 Scans for one spectrum were co–added at a spectral resolution of 4 cm^{-1}. The courses of the reactions were monitored using TLC aluminium sheets with Silica gel 60 F254 (Merck). The column chromatography was performed on Silica gel 60 (Merck). HPLC was performed on Büchi Pure 850 FlashPrep chromatography instrument using Prontosil, 150 × 20 mm, 5 µm column.

3.2. Synthetic Procedures

3.2.1. Quinoline Derivative **4a**

Calixarene **2** (0.122 g, 0.20 mmol) was dissolved in 5 mL of toluene at room temperature. Acetaldehyde (0.020 mL, 0.36 mmol) was added, and the resulting solution was stirred for 10 min. Then, trifluoroacetic acid (0.13 mL) was added and the colour of the solution immediately turned red. The reaction mixture was heated to 80 °C and stirred for 17 h. The solution was quenched by saturated $NaHCO_3$ (5 mL). The organic phase was separated, washed with water (2 × 20 mL) and dried over magnesium sulphate. The solvent was removed under reduced pressure to yield a crude product, which was further purified by thin-layer chromatography on silica gel (cyclohexane:ethyl acetate 20:1, v/v) to give the title compound **4a** as a yellow amorphous solid (0.061 g, 46%), m.p. 173–176 °C.

^{1}H NMR (CDCl$_3$, 400 MHz, 298 K) δ 8.00 (d, 1H, J = 8.6 Hz, Ar-*H*), 7.51 (s, 1H, Ar-*H*), 7.20 (d, 1H, J = 8.2 Hz, Ar-*H*), 7.16–7.08 (m, 2H, Ar-*H*), 6.92 (t, 1H, J = 7.4 Hz, Ar-*H*), 6.22–6.05 (m, 5H, Ar-*H*), 5.96–5.89 (m, 1H, Ar-*H*), 5.03 (d, 1H, J = 13.3 Hz, Ar-*CH$_2$*-Ar), 4.66 (d, 1H, J = 13.3 Hz, Ar-*CH$_2$*-Ar), 4.52 (d, 1H, J = 13.3 Hz, Ar-*CH$_2$*-Ar), 4.51 (d, 1H, J = 13.3 Hz, Ar-*CH$_2$*-Ar), 4.35 (d, 1H, J = 13.3 Hz, Ar-*CH$_2$*-Ar), 4.25–4.13 (m, 2H, O-*CH$_2$*), 4.13–4.03 (m, 2H, O-*CH$_2$*), 3.90–3.80 (m, 1H, O-*CH$_2$*), 3.79–3.69 (m, 3H, O-*CH$_2$*), 3.37 (d, 1H, J = 13.3 Hz, Ar-*CH$_2$*-Ar), 3.24–3.14 (m, 2H, Ar-*CH$_2$*-Ar), 2.76 (s, 3H, Ar-*CH$_3$*), 2.11–1.87 (m, 8H, O-CH$_2$-*CH$_2$*), 1.21–1.09 (m, 6H, O-CH$_2$-CH$_2$-*CH$_3$*), 0.99–0.87 (m, 6H, O-CH$_2$-CH$_2$-*CH$_3$*) ppm. ^{13}C NMR (CDCl$_3$, 100 MHz, 298 K) δ 159.1, 158.1, 157.3, 155.3 (2×), 147.4, 137.4, 137.3, 137.0, 135.6, 134.4, 133.5, 133.0, 132.8, 132.4, 128.9, 128.8, 127.6, 127.4, 127.2, 126.9, 125.5, 122.9, 122.1, 122.0, 121.7, 119.8, 77.1, 76.9, 76.8, 76.5, 31.5, 31.1, 31.0, 26.9, 25.6, 23.6, 23.1, 23.0, 22.8, 10.9 (2×), 9.9 (2×) ppm. IR (KBr) v 2961.1, 2931.8, 2874.3, 1590.4, 1454.9, 1383.1, 1245.0, 1193.7, 1005.9 cm^{-1}. HRMS (ESI$^+$) calcd for C$_{44}$H$_{51}$NO$_4$ 658.3891 [M+H]$^+$, 680.3710 [M+Na]$^+$, found *m/z* 658.3892 [M+H]$^+$ (100%), 680.3705 [M+Na]$^+$ (10%).

3.2.2. Quinoline Derivative **4b**

Calixarene **2** (0.101 g, 0.17 mmol) was dissolved in 5 mL of toluene at room temperature. Bromoacetaldehyde diethyl acetal (0.080 mL, 0.53 mmol) was added, and the resulting solution was stirred for 10 min. Then, trifluoroacetic acid (0.25 mL) was added and the colour of the solution immediately turned red. The reaction mixture was heated to 80 °C and stirred for 17 h. The solution was quenched by saturated $NaHCO_3$ (5 mL). The organic phase was separated, washed with water (2 × 20 mL) and dried over magnesium sulphate. The solvent was removed under reduced pressure to yield a crude product, which was further purified by thin-layer chromatography on silica gel (cyclohexane:ethyl acetate 6:1, v/v) to give the title compound **5b** as a brown amorphous solid (0.061 g, 37%), m.p. 179–182 °C.

^{1}H NMR (CDCl$_3$, 400 MHz, 298 K) δ 8.24 (s, 1H, Ar-*H*), 7.41 (s, 1H, Ar-*H*), 6.98 (dd, 2H, *J* = 7.4, 1.2 Hz, Ar-*H*), 6.93 (dd, 1H, *J* = 7.4, 1.2 Hz, Ar-*H*), 6.76 (t, 1H, *J* = 7.4 Hz, Ar-*H*), 6.25–6.16 (m, 4H, Ar-*H*), 6.11 (dd, 1H, *J* = 6.3, 2.7 Hz, Ar-*H*), 6.02 (dd, 1H, *J* = 6.3, 2.7 Hz, Ar-*H*), 5.30 (s, 1H, CH$_2$-*OH*), 4.89 (s, 2H, Ar-*CH$_2$*-OH), 4.74 (d, 1H, *J* = 13.7 Hz, Ar-*CH$_2$*-Ar), 4.65 (d, 1H, *J* = 13.3 Hz, Ar-*CH$_2$*-Ar), 4.47 (d, 1H, *J* = 13.3 Hz, Ar-*CH$_2$*-Ar), 4.46 (d, 1H, *J* = 13.3 Hz, Ar-*CH$_2$*-Ar), 4.42 (d, 1H, *J* = 13.7 Hz, Ar-*CH$_2$*-Ar), 4.15 (dd, 1H, *J* = 8.6, 7.0 Hz, O-*CH$_2$*), 4.07–3.95 (m, 2H, O-*CH$_2$*), 3.87–3.66 (m, 5H, O-*CH$_2$*), 3.37 (d, 1H, *J* = 13.3 Hz, Ar-*CH$_2$*-Ar), 3.16 (d, 1H, *J* = 13.3 Hz, Ar-*CH$_2$*-Ar), 3.15 (d, 1H, *J* = 13.7 Hz, Ar-*CH$_2$*-Ar), 2.06–1.96 (m, 3H, O-CH$_2$-*CH$_2$*), 1.95–1.84 (m, 5H, O-CH$_2$-*CH$_2$*), 1.10 (t, 6H, *J* = 7.4 Hz, O-CH$_2$-CH$_2$-*CH$_3$*), 0.93 (t, 3H, *J* = 7.4 Hz, O-CH$_2$-CH$_2$-*CH$_3$*), (t, 3H, *J* = 7.4 Hz, O-CH$_2$-CH$_2$-*CH$_3$*) ppm. ^{13}C NMR (CDCl$_3$, 125 MHz, 298 K) δ 159.6, 157.7, 155.5 (2×), 154.2, 144.2, 139.2, 138.7, 136.9, 136.5, 134.1, 133.7, 133.5, 132.5, 132.2, 128.7 (2×), 128.0, 127.5, 127.4, 127.3, 125.3, 124.9, 122.1 (2×), 121.8, 113.2, 77.0 (2×), 76.9, 76.5, 63.7, 31.5, 31.1, 31.0, 23.7, 23.5, 23.4, 23.2, 23.1, 10.8 (2×), 10.00 (2×) ppm. IR (KBr) ν 2961.6, 2931.8, 2874.7, 1729.8, 1588.2, 1455.7, 1383.8, 1203.5, 1086.9 cm^{-1}. HRMS (ESI$^+$) calcd for C$_{44}$H$_{50}$NBrO$_5$ 776.2744 [M+Na]$^+$, 792.2484 [M+K]$^+$, found *m/z* 776.2748 [M+Na]$^+$ (100%), 792.2476 [M+K]$^+$ (5%).

3.2.3. Quinoline Derivative **6a**

Calixarene **3** (0.101 g, 0.14 mmol) was dissolved in 5 mL of toluene at room temperature. Acetaldehyde (0.020 mL, 0.36 mmol) was added, and the resulting solution was stirred for 10 min. Then, trifluoroacetic acid (0.11 mL) was added and the colour of the solution immediately turned red. The reaction mixture was heated to 80 °C and stirred for 15 h. The solution was quenched by saturated NaHCO$_3$ (5 mL). The organic phase was separated, washed with water (2 × 20 mL) and dried over magnesium sulphate. The solvent was removed under reduced pressure to yield a crude product, which was further purified by thin-layer chromatography on silica gel (cyclohexane:ethyl acetate 3:1, *v/v*) to give the title compound **6a** as a brown amorphous solid (0.019 g, 18%), m.p. 170–173 °C.

^{1}H NMR (CDCl$_3$, 400 MHz, 298 K) δ 8.35 (d, 1H, *J* = 8.6 Hz, Ar-*H*), 7.79 (s, 1H, Ar-*H*), 7.26 (d, 1H, *J* = 9.0 Hz, Ar-*H*), 7.01 (dd, 1H, *J* = 7.4, 1.6 Hz, Ar-*H*), 6.94 (dd, 1H, *J* = 7.4, 1.2 Hz, Ar-*H*), 6.76 (t, 1H, *J* = 7.4 Hz, Ar-*H*), 6.22–6.13 (m, 5H, Ar-*H*), 5.94 (dd, 1H, *J* = 7.0, 1.6 Hz, Ar-*H*), 4.63 (d, 1H, *J* = 12.9 Hz, Ar-*CH$_2$*-Ar), 4.60 (d, 1H, *J* = 13.7 Hz, Ar-*CH$_2$*-Ar), 4.46 (d, 1H, *J* = 13.3 Hz, Ar-*CH$_2$*-Ar), 4.46 (d, 1H, *J* = 13.3 Hz, Ar-*CH$_2$*-Ar), 4.14 (dt, 1H, *J* = 10.6, 5.5 Hz, O-*CH$_2$*), 4.05–3.95 (m, 3H, O-*CH$_2$*), 3.92 (d, 1H, *J* = 14.1 Hz, Ar-*CH$_2$*-Ar), 3.83–3.70 (m, 4H, O-*CH$_2$*), 3.42 (d, 1H, *J* = 13.3 Hz, Ar-*CH$_2$*-Ar), 3.19–3.13 (m, 2H, Ar-*CH$_2$*-Ar), 2.75 (s, 3H, Ar-*CH$_3$*), 2.07–1.86 (m, 8H, O-CH$_2$-*CH$_2$*), 1.11 (t, 3H, *J* = 7.4 Hz, O-CH$_2$-CH$_2$-*CH$_3$*), 1.10 (t, 3H, *J* = 7.4 Hz, O-CH$_2$-CH$_2$-*CH$_3$*), 0.96–0.89 (m, 6H, O-CH$_2$-CH$_2$-*CH$_3$*) ppm. ^{13}C NMR (CDCl$_3$, 100 MHz, 298 K) δ 157.7, 156.1, 155.6, 155.5, 155.4 (2×), 136.8, 136.7 (2×), 133.7 (2×), 133.2, 132.3, 130.3, 128.8, 128.6 (2×), 127.7 (2×), 127.6, 126.8, 125.8, 122.1 (3×), 121.7, 121.0, 76.9 (3×), 76.4, 31.7, 31.1, 30.9, 24.2, 23.5 (2×), 23.4, 23.1, 23.0, 10.8, 10.7, 10.0, 9.9 ppm. IR (KBr) ν 2960.5, 2920.4, 2874.1, 1455.4, 1383.7, 1208.3, 1086.9 cm^{-1}. HRMS (ESI$^+$) calcd for C$_{44}$H$_{51}$NO$_4$ 658.3891 [M+H]$^+$, 696.3450 [M+Na]$^+$, found *m/z* 658.3894 [M+H]$^+$ (100%), 696.3443 [M+Na]$^+$ (3%).

3.2.4. Bis-Quinoline Derivative **14**

Calixarene **7** (0.104 g, 0.17 mmol) was dissolved in 5 mL of toluene at room temperature. Acetaldehyde (0.040 mL, 0.71 mmol) was added, and the resulting solution was stirred for 10 min. Then, trifluoroacetic acid (0.22 mL) was added and the colour of the solution immediately turned red. The reaction mixture was heated to 80 °C and stirred for 15 h. The solution was quenched by saturated NaHCO$_3$ (5 mL). The organic phase was separated, washed with water (2 × 20 mL) and dried over magnesium sulphate. The solvent was removed under reduced pressure to yield a crude product which was further purified by thin-layer chromatography on silica gel (cyclohexane:ethyl acetate 6:1, *v/v*) to give the title compound **14** as a yellow amorphous solid (0.061 g, 37%), m.p. 185–188 °C.

^{1}H NMR (CDCl$_3$, 400 MHz, 298 K) δ 8.00 (d, 2H, J = 8.2 Hz, Ar-*H*), 7.52 (s, 2H, Ar-*H*), 7.20 (d, 2H, J = 8.2 Hz, Ar-*H*), 6.21–6.05 (m, 2H, Ar-*H*), 6.03–5.95 (m, 2H, Ar-*H*), 5.84 (dd, 2H, J = 6.7, 2.0 Hz, Ar-*H*), 5.04 (d, 2H, J = 13.3 Hz, Ar-*CH$_2$*-Ar), 4.67 (d, 2H, J = 13.3 Hz, Ar-*CH$_2$*-Ar), 4.35 (d, 2H, J = 13.3 Hz, Ar-*CH$_2$*-Ar), 4.25–4.16 (m, 2H, O-*CH$_2$*), 3.90–3.68 (m, 6H, O-*CH$_2$*), 3.38 (d, 2H, J = 13.3 Hz, Ar-*CH$_2$*-Ar), 2.75 (s, 6H, Ar-*CH$_3$*), 2.13–1.85 (m, 8H, O-CH$_2$-*CH$_2$*), 1.17 (t, 6H, J = 7.4 Hz, O-CH$_2$-CH$_2$-*CH$_3$*), 0.91 (t, 6H, J = 7.4 Hz, O-CH$_2$-CH$_2$-*CH$_3$*) ppm. ^{13}C NMR (CDCl$_3$, 100 MHz, 298 K) δ 159.2, 157.3, 155.4, 147.4, 137.6, 135.6, 134.7, 133.0, 131.8, 127.5, 126.7, 125.5, 122.9, 122.1, 119.8, 76.9 (2×), 31.5, 25.6, 23.6, 23.1, 22.6, 10.9, 9.9 ppm. IR (KBr) ν 2960.1, 2921.7, 2874.2, 1602.4, 1492.0, 1453.8, 1382.6, 1184.0, 1085.3 cm^{-1}. HRMS (ESI$^+$) calcd for C$_{48}$H$_{54}$N$_2$O$_4$ 723.4156 [M+H]$^+$, 745.3976 [M+Na]$^+$, found *m/z* 723.4156 [M+H]$^+$ (100%), 745.3972 [M+Na]$^+$ (25%).

3.2.5. Quinoline Derivatives **15a** and **15b**

Calixarene **8** (0.119 g, 0.19 mmol) was dissolved in 5 mL of toluene at room temperature. Acetaldehyde (0.040 mL, 0.71 mmol) was added, and the resulting solution was stirred for 10 min. Then, trifluoroacetic acid (0.25 mL) was added and the colour of the solution immediately turned red. The reaction mixture was heated to 80 °C and stirred for 17 h. The solution was quenched by saturated NaHCO$_3$ (5 mL). The organic phase was separated, washed with water (2 × 20 mL) and dried over magnesium sulphate. The solvent was removed under reduced pressure to yield a crude product which was further purified by thin-layer chromatography on silica gel (cyclohexane:ethyl acetate 4:1, *v/v*) to give an inseparable mixture of title compounds **15a** and **15b** as a yellow amorphous solid (0.020 g, 14%), m.p. 152–155 °C.

^{1}H NMR (CDCl$_3$, 400 MHz, 298 K) δ 8.39 (t, 2H, J = 9.0 Hz, Ar-*H*), 7.98 (d, 2H, J = 8.6 Hz, Ar-*H*), 7.87 (s, 2H, Ar-*H*), 7.51 (s, 2H, Ar-*H*), 7.29 (d, 2H, J = 9.0 Hz, Ar-*H*), 7.19 (d, 2H, J = 8.2 Hz, Ar-*H*), 6.09–5.95 (m, 7H, Ar-*H*), 5.93 (t, 1H, J = 7.4 Hz, Ar-*H*), 5.86 (dd, 1H, J = 7.4, 1.6 Hz, Ar-*H*), 5.81 (dd, 1H, J = 7.4, 1.6 Hz, Ar-*H*), 5.78–5.71 (m, 2H, Ar-*H*), 5.03 (d, 1H, J = 13.3 Hz, Ar-*CH$_2$*-Ar), 5.02 (d, 1H, J = 13.7 Hz, Ar-*CH$_2$*-Ar), 4.70–4.55 (m, 4H, Ar-*CH$_2$*-Ar), 4.33 (d, 1H, J = 12.9 Hz, Ar-*CH$_2$*-Ar), 4.31 (d, 1H, J = 13.3 Hz, Ar-*CH$_2$*-Ar), 4.26–4.00 (m, 8H, O-*CH$_2$*), 3.94 (d, 1H, J = 14.1 Hz, Ar-*CH$_2$*-Ar), 3.94 (d, 1H, J = 13.3 Hz, Ar-*CH$_2$*-Ar), 3.87–3.68 (m, 8H, O-*CH$_2$*), 3.45 (d, 1H, J = 13.3 Hz, Ar-*CH$_2$*-Ar), 3.44 (d, 1H, J = 12.9 Hz, Ar-*CH$_2$*-Ar), 3.36 (d, 1H, J = 13.3 Hz, Ar-*CH$_2$*-Ar), 3.35 (d, 1H, J = 13.3 Hz, Ar-*CH$_2$*-Ar), 2.76 (s, 3H, Ar-*CH$_3$*), 2.73 (s, 3H, Ar-*CH$_3$*), 2.14–1.83 (m, 16H, O-CH$_2$-*CH$_2$*), 1.19–1.10 (m, 12H, O-CH$_2$-CH$_2$-*CH$_3$*), 0.97–0.84 (m, 12H, O-CH$_2$-CH$_2$-*CH$_3$*) ppm. HRMS (ESI$^+$) calcd for C$_{48}$H$_{54}$N$_2$O$_4$ 723.4156 [M+H]$^+$, 745.3976 [M+Na]$^+$, found *m/z* 723.4162 [M+H]$^+$ (100%), 745.3974 [M+Na]$^+$ (20%).

3.2.6. Quinoline Derivative **15c**

Calixarene **15c** was isolated from the same reaction mixture as compounds **15a** and **15b**. The product was obtained as a yellow amorphous solid (0.014 g, 10%), m.p. 163–166 °C.

^{1}H NMR (CDCl$_3$, 500 MHz, 298 K) δ 7.98 (d, 1H, J = 8.2 Hz, Ar-*H*), 7.48 (s, 1H, Ar-*H*), 7.18 (d, 1H, J = 8.2 Hz, Ar-*H*), 6.43 (s, 1H, Ar-*NH*-CH$_2$), 6.26–6.01 (m, 5H, Ar-*H*), 5.89 (d, 1H, J = 7.0 Hz, Ar-*H*), 4.99 (d, 1H, J = 13.3 Hz, Ar-*CH$_2$*-Ar), 4.62 (d, 1H, J = 13.3 Hz, Ar-*CH$_2$*-Ar), 4.43 (d, 1H, J = 13.3 Hz, Ar-*CH$_2$*-Ar), 4.42 (d, 1H, J = 12.9 Hz, Ar-*CH$_2$*-Ar), 4.21–4.08 (m, 2H, O-*CH$_2$*), 3.98–3.89 (m, 2H, O-*CH$_2$*), 3.83–3.65 (m, 4H, O-*CH$_2$*), 3.34 (d, 1H, J = 13.3 Hz, Ar-*CH$_2$*-Ar), 3.16 (q, 2H, J = 7.0 Hz, NH-*CH$_2$*-CH$_3$), 3.05 (d, 2H, J = 13.3 Hz, Ar-*CH$_2$*-Ar), 2.74 (s, 3H, Ar-*CH$_3$*), 2.10–1.83 (m, 8H, O-CH$_2$-*CH$_2$*), 1.29 (t, 3H, J = 7.0 Hz, NH-CH$_2$-*CH$_3$*), 1.12 (t, 3H, J = 7.4 Hz, O-CH$_2$-CH$_2$-*CH$_3$*), 1.11 (t, 3H, J = 7.4 Hz, O-CH$_2$-CH$_2$-*CH$_3$*), 0.90 (t, 3H, J = 7.4 Hz, O-CH$_2$-CH$_2$-*CH$_3$*), 0.87 (t, 3H, J = 7.4 Hz, O-CH$_2$-CH$_2$-*CH$_3$*) ppm. ^{13}C NMR (CDCl$_3$, 125 MHz, 298 K) δ 157.2, 155.3, 155.2, 137.7, 137.3, 135.6, 134.3, 133.5, 132.8, 132.3, 127.6, 127.3, 127.1, 126.8, 125.4, 122.8, 122.0, 121.9, 119.7, 77.0, 76.9, 76.7, 76.6, 31.5, 31.2, 31.1, 29.7, 25.5, 23.6, 23.5, 23.0, 22.9, 15.0, 10.9, 10.8, 9.9 (2×) ppm. IR (KBr) ν 2959.7, 1219.1, 173.9, 1958.3, 1604.5, 1454.2, 1383.3, 1087.7 cm^{-1}. HRMS (ESI$^+$) calcd for C$_{46}$H$_{56}$N$_2$O$_4$ 701.4313 [M+H]$^+$, 723.4132 [M+Na]$^+$, found *m/z* 701.4318 [M+H]$^+$ (100%), 723.4134 [M+Na]$^+$ (75%).

3.2.7. Quinoline Derivative **16**

Calixarene **12** (0.101 g, 0.14 mmol) was dissolved in 5 mL of toluene at room temperature. Acetaldehyde (0.020 mL, 0.36 mmol) was added, and the resulting solution was stirred for 10 min. Then, trifluoroacetic acid (0.09 mL) was added and the colour of the solution immediately turned red. The reaction mixture was heated to 80 °C and stirred for 17 h. The solution was quenched by saturated $NaHCO_3$ (5 mL). The organic phase was separated, washed with water (2 × 20 mL) and dried over magnesium sulphate. The solvent was removed under reduced pressure to yield crude product, which was further purified by thin-layer chromatography on silica gel (cyclohexane:ethyl acetate 4:1, v/v) to give the title compound **16** as a yellow amorphous solid (0.051 g, 47%), m.p. 163–166 °C.

^{1}H NMR ($CDCl_3$, 400 MHz, 298 K) δ 7.92 (d, 1H, J = 8.2 Hz, Ar-*H*), 7.72 (br s, 1H, Ar-*H*), 7.40 (s, 1H, Ar-*H*), 7.14 (d, 1H, J = 8.2 Hz, Ar-*H*), 7.09 (br s, 1H, Ar-*H*), 6.32–6.07 (m, 6H, Ar-*H*), 4.92 (d, 1H, J = 12.9 Hz, Ar-*CH$_2$*-Ar), 4.62 (d, 1H, J = 13.3 Hz, Ar-*CH$_2$*-Ar), 4.53–4.42 (m, 1H, Ar-*CH$_2$*-Ar), 4.34 (d, 1H, J = 12.9 Hz, Ar-*CH$_2$*-Ar), 4.17–3.93 (m, 4H, O-*CH$_2$*), 3.89–3.79 (m, 1H, O-*CH$_2$*), 3.79–3.67 (m, 3H, O-*CH$_2$*), 3.35 (d, 1H, J = 13.3 Hz, Ar-*CH$_2$*-Ar), 3.17 (d, 1H, J = 13.3 Hz, Ar-*CH$_2$*-Ar), 3.15 (d, 1H, J = 13.3 Hz, Ar-*CH$_2$*-Ar), 2.72 (s, 3H, Ar-*CH$_3$*), 2.07–1.82 (m, 8H, O-*CH$_2$*-*CH$_2$*), 1.18–1.03 (m, 6H, O-*CH$_2$*-*CH$_2$*-*CH$_3$*), 1.00–0.85 (m, 6H, O-*CH$_2$*-*CH$_2$*-*CH$_3$*) ppm. ^{13}C NMR ($CDCl_3$, 100 MHz, 298 K) δ 158.8, 158.7, 157.1, 156.0, 155.5 (2×), 137.9, 137.5, 137.2, 136.9, 135.4, 134.8, 134.5, 133.0 (2×), 132.2, 128.6, 128.3, 127.7, 127.6, 126.9, 125.6, 122.9, 122.2, 121.9, 120.8 (2×), 119.7, 117.3, 77.0, 76.9 (2×), 76.1, 31.3, 31.0 (2×), 25.5, 23.5 (3×), 23.1, 23.0, 10.7, 10.0 (2×), 9.6 ppm. IR (KBr) v 3305.1, 2962.0, 2933.1, 2875.5, 1710.2, 1605.4, 1455.9, 1183.1, 758.4 cm^{-1}. HRMS (ESI$^+$) calcd for $C_{46}H_{51}F_3N_2O_5$ 769.3823 [M+H]$^+$, 791.3642 [M+Na]$^+$, 807.3382 [M+K]$^+$, found m/z 769.3827 [M+H]$^+$ (100%), 791.3638 [M+Na]$^+$ (30%), 807.3375 [M+K]$^+$ (18%).

3.2.8. Quinoline Derivative **17a**

Calixarene **13** (0.124 g, 0.17 mmol) was dissolved in 5 mL of toluene at room temperature. Acetaldehyde (0.020 mL, 0.36 mmol) was added, and the resulting solution was stirred for 10 min. Then, trifluoroacetic acid (0.11 mL) was added and the colour of the solution immediately turned red. The reaction mixture was heated to 80 °C and stirred for 17 h. The solution was quenched by saturated $NaHCO_3$ (5 mL). The organic phase was separated, washed with water (2 × 20 mL) and dried over magnesium sulphate. The solvent was removed under reduced pressure to yield a crude product, which was further purified by preparative HPLC to give the title compounds 17a as a yellow amorphous solid (0.023 g, 17%), m.p. 120–123 °C. In the ^{13}C NMR spectrum, the number of signals does not correspond to the number of carbons due to the low intensity of certain signals.

^{1}H NMR ($CDCl_3$, 500 MHz, 298 K) δ 7.73 (s, 1H, Ar-*H*), 7.23–6.73 (m, 8H, Ar-*H*, Ar-*NH*-COCF$_3$), 6.67–6.53 (m, 1H, Ar-*H*), 6.26–6.05 (m, 2H, Ar-*H*), 4.84 (d, 1H, J = 14.0 Hz, Ar-*CH$_2$*-Ar), 4.61 (d, 1H, J = 13.5 Hz, Ar-*CH$_2$*-Ar), 4.47 (d, 1H, J = 14.9 Hz, Ar-*CH$_2$*-Ar), 4.40 (d, 1H, J = 13.5 Hz, Ar-*CH$_2$*-Ar), 4.01–3.90 (m, 4H, O-*CH$_2$*), 3.80 (d, 1H, J = 14.6 Hz, Ar-*CH$_2$*-Ar), 3.77–3.59 (m, 4H, O-*CH$_2$*), 3.37 (d, 1H, J = 13.5 Hz, Ar-*CH$_2$*-Ar), 3.19–3.05 (m, 2H, Ar-*CH$_2$*-Ar), 2.52 (s, 3H, Ar-*CH$_3$*), 1.98–1.74 (m, 8H, O-*CH$_2$*-*CH$_2$*), 1.10 (t, 3H, J = 7.1 Hz, O-*CH$_2$*-*CH$_2$*-*CH$_3$*), 1.04 (t, 3H, J = 7.4 Hz, O-*CH$_2$*-*CH$_2$*-*CH$_3$*), 0.92 (t, 3H, J = 7.1 Hz, O-*CH$_2$*-*CH$_2$*-*CH$_3$*), 1.06–0.81 (m, 3H, O-*CH$_2$*-*CH$_2$*-*CH$_3$*) ppm. ^{13}C NMR ($CDCl_3$, 125 MHz, 298 K) δ 155.7, 145.1, 132.4, 129.8, 129.1, 129.0, 128.6, 128.4, 122.6, 122.3, 121.9, 118.2, 114.9, 114.1, 77.2, 77.1, 76.9, 76.3, 31.1, 30.9, 29.7, 26.9, 24.6, 23.5, 23.4, 23.1, 21.9, 10.7 (2×), 10.0, 9.6. ppm. IR (KBr) v 3360.6, 2961.8, 2920.2, 2875.8, 1729.1, 1454.9, 1203.2, 1157.5 cm^{-1}. HRMS (ESI$^+$) calcd for $C_{46}H_{51}F_3N_2O_5$ 769.3823 [M+H]$^+$, 791.3642 [M+Na]$^+$, 807.3382 [M+K]$^+$, found m/z 769.3827 [M+H]$^+$ (100%), 791.3638 [M+Na]$^+$ (30%), 807.3375 [M+K]$^+$ (15%).

3.2.9. Quinoline Derivative **17b**

Calixarene **17b** was isolated from the same reaction mixture as compound **17a**. This product was isolated as a yellow amorphous solid (0.020 g, 15%), m.p. 122–125 °C. In the

^{13}C NMR spectrum, the number of signals does not correspond to the number of carbons due to the low intensity of certain signals.

^{1}H NMR (CDCl$_3$, 400 MHz, 298 K) δ 8.03 (s, 1H, Ar-*H*), 7.77 (s, 1H, Ar-*NH*-COCF$_3$), 7.06–6.99 (m, 2H, Ar-*H*), 6.83–6.48 (m, 6H, Ar-*H*), 6.29–5.96 (m, 2H, Ar-*H*), 4.73 (d, 1H, *J* = 14.1 Hz, Ar-*CH$_2$*-Ar), 4.60 (d, 1H, *J* = 13.7 Hz, Ar-*CH$_2$*-Ar), 4.44 (d, 1H, *J* = 14.4 Hz, Ar-*CH$_2$*-Ar), 4.38 (d, 1H, *J* = 13.8 Hz, Ar-*CH$_2$*-Ar), 4.04–3.90 (m, 4H, O-*CH$_2$*), 3.88–3.72 (m, 5H, O-*CH$_2$*, Ar-*CH$_2$*-Ar), 3.39 (d, 1H, *J* = 13.8 Hz, Ar-*CH$_2$*-Ar), 3.24 (d, 1H, *J* = 14.5 Hz, Ar-*CH$_2$*-Ar), 3.10 (d, 1H, *J* = 13.8 Hz, Ar-*CH$_2$*-Ar), 2.60 (s, 3H, Ar-*CH$_3$*), 2.02–1.77 (m, 8H, O-CH$_2$-*CH$_2$*), 1.08–0.92 (m, 12H, O-CH$_2$-CH$_2$-*CH$_3$*) ppm. ^{13}C NMR (CDCl$_3$, 100 MHz, 298 K) δ 156.0, 144.9, 140.2, 132.8, 130.4, 128.9, 128.3, 127.9, 124.8, 122.4, 121.9, 120.2, 118.7, 117.2, 77.0, 76.8 (2×), 31.7, 31.1, 29.7, 26.9, 24.8, 23.3, 23.2, 22.9, 10.5, 10.4, 10.2 (2×) ppm. IR (KBr) ν 3352.4, 2961.7, 2920.3, 2875.9, 1728.4, 1454.4, 1203.9, 1157.9 cm^{-1}. HRMS (ESI$^+$) calcd for C$_{46}$H$_{51}$F$_3$N$_2$O$_5$ 769.3823 [M+H]$^+$, 791.3642 [M+Na]$^+$, 807.3382 [M+K]$^+$, found *m/z* 769.3815 [M+H]$^+$ (100%), 791.3635 [M+Na]$^+$ (8%), 807.3375 [M+K]$^+$ (20%).

3.3. Chiral Separation

Chiral separation of **6a** was performed using a Büchi Pure 850 FlashPrep chromatography instrument consisting of a binary pump module, UV-vis detector, column manager and fraction collector. The suitable conditions allowing for efficient enantioseparation were first proven on the analytical scale using chiral polysaccharide column ChiralArt Amylose-SA (250 × 4.6 mm ID, 5 µm). In preparative mode, a polysaccharide column ChiralArt Amylose-SA (250 × 20 mm ID, 5 µm) was employed using cyclohexane/DCM: 92/8 *v/v* as a mobile phase.

3.4. NMR Titrations

In each case, calixarene was dissolved in a specified amount of CDCl$_3$ or C$_2$D$_2$Cl$_4$ and 0.5 mL of calixarene solution was put in an NMR tube. A specific amount of guest was added to the calixarene solution (0.6 mL), and the aliquots of guest were gradually added to the NMR tube to achieve different calixarene/guest ratios, ensuring constant calixarene concentration during the experiment. The complexation constants were determined by analyzing CIS (complexation induced chemical shifts) of calixarene protons using a nonlinear curve-fitting procedure (program BindFit) [30].

3.5. X-ray Measurements

Crystallographic data for **16**. M = 800.94 g·mol^{-1}, triclinic system, space group *P-1*, a = 15.52150 (2) Å, b = 18.08924 (3) Å, c = 18.11720 (2) Å, α = 77.920 (3)°, β = 71.014 (2)°, γ = 64.684 (2)° Z = 4, V = 4333.46 (9) Å^3, D_c = 1.228 g·cm^{-3}, μ(Cu-Kα) = 0.73 mm^{-1}, crystal dimensions of 0.41 × 0.29 × 0.26 mm. Data were collected at 200 (2) K on a Bruker D8 Venture Photon CMOS diffractometer with Incoatec microfocus sealed tube Cu-Kα radiation. The structure was solved by charge flipping methods [31] and anisotropically refined by full matrix least squares on F squared using the CRYSTALS [32] to final value R = 0.085 and wR = 0.242 using 15,813 independent reflections (θ_{max} = 68.4°), 1370 parameters and 481 restrains. The hydrogen atoms bonded to carbon atoms were placed in calculated positions refined with riding constrains, while hydrogen atoms bonded to oxygen and nitrogen were refined using soft restraints. The disordered functional group positions were found in difference electron density maps and refined with restrained geometry. MCE [33] was used for visualization of electron density maps. The occupancy of the disordered functional group was fully constrained. The structure was deposited into the Cambridge Structural Database under the number CCDC 2215763.

4. Conclusions

In summary, the construction of inherently chiral calixarenes by the intramolecular cyclization of suitable intermediates suffers from a limited number of suitable substrates for these reactions. Here, we report on an easy way to prepare one class of such compounds: calixquinolines, which can be obtained by the reaction of aldehydes with easily accessible aminocalix[4]arenes under acidic conditions (Doebner–Miller reaction). The synthetic procedure represents a very straightforward approach to the inherently chiral macrocyclic systems. The complexation studies revealed the ability of these compounds to complex quaternary ammonium salts with different stoichiometries depending on the guest molecules; moreover, we demonstrated their ability to carry out the enantioselective complexation of chiral *N*-methylammonium salts.

Supplementary Materials: The following supporting information can be downloaded at: https://www.mdpi.com/article/10.3390/molecules27238545/s1. Spectral characterization of all new compounds (^{1}H NMR, ^{13}C NMR, HRMS, IR) and the NMR complexation studies.

Author Contributions: Conceptualization, writing—review and editing, P.L.; experimental work, synthesis, spectra analysis, editing, M.T.; NMR measurement and VT NMR experiments, H.D.; X-ray measurement, V.E. All authors have read and agreed to the published version of the manuscript.

Funding: This research was funded by the Czech Science Foundation, grant number 20-07833S.

Institutional Review Board Statement: Not applicable.

Informed Consent Statement: Not applicable.

Data Availability Statement: All experimental data are provided in the Supplementary information.

Conflicts of Interest: The authors declare no conflict of interest.

References

1. Gutsche, C.D. *Calixarenes: An Introduction*; RSC Publishing: Cambridge, UK, 2008.
2. Mandolini, L.; Ungaro, R. *Calixarenes in Action*; Imperial College Press: London, UK, 2000.
3. Vicens, J.; Harrowfield, J.; Baklouti, L. *Calixarenes in the Nanoworld*; Springer: Dordrecht, The Netherlands, 2007.
4. Neri, P.; Sessler, J.L.; Wang, M.X. *Calixarenes and Beyond*; Springer: Cham, Switzerland, 2016.
5. Leray, I.; Valeur, B. Calixarene-Based Fluorescent Molecular Sensors for Toxic Metals. *Eur. J. Inorg. Chem.* **2009**, *24*, 3525–3535. [CrossRef]
6. Siddiqui, S.; Cragg, P.J. Design and Synthesis of Transition Metal and Inner Transition Metal Binding Calixarenes. *Mini-Rev. Org. Chem.* **2009**, *6*, 283–299. [CrossRef]
7. Mutihac, L.; Buschmann, H.-J.; Mutihac, R.-C.; Schollmeyer, E. Complexation and separation of amines, amino acids, and peptides by functionalized calix[n]arenes. *J. Incl. Phenom. Macrocyclic Chem.* **2005**, *51*, 1–10. [CrossRef]
8. Lhotak, P. Anion receptors based on calixarenes. *Top. Curr. Chem.* **2005**, *255*, 65–95.
9. Kongor, A.R.; Mehta, V.A.; Modi, K.M.; Panchal, M.K.; Dey, S.A.; Panchal, U.S.; Jain, V.K. Calix-Based Nanoparticles: A Review. *Top. Curr. Chem.* **2016**, *374*, 1–46. [CrossRef]
10. Szumna, A. Inherently chiral concave molecules-from synthesis to applications. *Chem. Soc. Rev.* **2010**, *39*, 4274–4285. [CrossRef] [PubMed]
11. Arnott, G.E. Inherently Chiral Calixarenes: Synthesis and Applications. *Chem.-Eur. J.* **2018**, *24*, 1744–1754. [CrossRef] [PubMed]
12. Li, S.-Y.; Xu, Y.-W.; Liu, Y.-W.; Su, C.-Y. Inherently chiral calixarenes. Synthesis, optical resolution, chiral recognition and asymmetric catalysis. *Int. J. Mol. Sci.* **2011**, *12*, 429–455. [CrossRef]
13. Lhoták, P. Direct meta substitution of calix[4]arenes. *Org. Biomol. Chem.* **2022**, *20*, 7377–7390. [CrossRef]
14. Slavik, P.; Dudic, M.; Flidrova, K.; Sykora, J.; Cisarova, I.; Bohm, S.; Lhotak, P. Unprecedented Meta-Substitution of Calixarenes: Direct Way to Inherently Chiral Derivatives. *Org. Lett.* **2012**, *14*, 3628–3631. [CrossRef]
15. Ikeda, A.; Yoshimura, M.; Lhotak, P.; Shinkai, S. Synthesis and optical resolution of naphthalene-containing inherently chiral calix[4] arenes derived by intramolecular ring closure or stapling of proximal phenyl units. *J. Chem. Soc. Perkin Trans.* **1996**, *1*, 1945–1950. [CrossRef]
16. Hueggenberg, W.; Seper, A.; Oppel, I.M.; Dyker, G. Multifold photocyclization reactions of styrylcalix[4]arenes. *Eur. J. Org. Chem.* **2010**, 6786–6797. [CrossRef]
17. Elaieb, F.; Semeril, D.; Matt, D.; Pfeffer, M.; Bouit, P.-A.; Hissler, M.; Gourlaouen, C.; Harrowfield, J. Calix[4]arene-fused phospholes. *Dalton Trans.* **2017**, *46*, 9833–9845. [CrossRef] [PubMed]
18. Tlusty, M.; Dvorakova, H.; Cejka, J.; Kohout, M.; Lhotak, P. Regioselective formation of the quinazoline moiety on the upper rim of calix[4]arene as a route to inherently chiral systems. *New J. Chem.* **2020**, *44*, 6490–6500. [CrossRef]
19. Miao, R.; Zheng, Q.-Y.; Chen, C.-F.; Huang, Z.-T. Efficient Syntheses and Resolutions of Inherently Chiral Calix[4]quinolines in the Cone and Partial-Cone Conformation. *J. Org. Chem.* **2005**, *70*, 7662–7671. [CrossRef] [PubMed]

20. Wang, T.; Ledeboer, M.W.; Duffy, J.P.; Pierce, A.C.; Zuccola, H.J.; Block, E.; Shlyakter, D.; Hogan, J.K.; Bennani, Y.L. A novel chemotype of kinase inhibitors: Discovery of 3,4-ring fused 7-azaindoles and deazapurines as potent JAK2 inhibitors. *Bioorg. Med. Chem. Lett.* **2010**, *20*, 153–156. [CrossRef] [PubMed]
21. Zheng, L.; Xiang, J.; Dang, Q.; Guo, S.; Bai, X. Design and synthesis of a tetracyclic pyrimidine-fused benzodiazepine library. *J. Comb. Chem.* **2006**, *8*, 381–387. [CrossRef]
22. Tlusty, M.; Slavik, P.; Kohout, M.; Eigner, V.; Lhotak, P. Inherently Chiral Upper-Rim-Bridged Calix[4]arenes Possessing a Seven Membered Ring. *Org. Lett.* **2017**, *19*, 2933–2936. [CrossRef]
23. Tlusty, M.; Eigner, V.; Babor, M.; Kohout, M.; Lhotak, P. Synthesis of upper rim-double-bridged calix[4]arenes bearing seven membered rings and related compounds. *RSC Adv.* **2019**, *9*, 22017–22030. [CrossRef]
24. Liska, A.; Flidrova, K.; Lhotak, P.; Ludvik, J. Influence of structure on electrochemical reduction of isomeric mono- and di-, nitro- or nitrosocalix[4]arenes. *Monatsh. Chem.* **2015**, *146*, 857–862. [CrossRef]
25. Wu, Y.-C.; Liu, L.; Li, H.-J.; Wang, D.; Chen, Y.-J. Skraup−Doebner−Von Miller Quinoline Synthesis Revisited: Reversal of the Regiochemistry for γ-Aryl-β,γ-unsaturated α-Ketoesters. *J. Org. Chem.* **2006**, *71*, 6592–6595. [CrossRef] [PubMed]
26. Su, L.-L.; Zheng, Y.-W.; Wang, W.-G.; Chen, B.; Wei, X.-Z.; Wu, L.-Z.; Tung, C.-H. Photocatalytic Synthesis of Quinolines via Povarov Reaction under Oxidant-Free Conditions. *Org. Lett.* **2022**, *24*, 1180–1185. [CrossRef] [PubMed]
27. Kelderman, E.; Verboom, W.; Engbersen, J.F.; Reinhoudt, D.N.; Heesink, G.J.; van Hulst, N.F.; Derhaeg, L.; Persoons, A. Nitrocalix[4] arenes as molecules for second-order nonlinear optics. *Angew. Chem. Int. Ed. Engl.* **1992**, *31*, 1075–1077. [CrossRef]
28. van Wageningen, A.M.; Snip, E.; Verboom, W.; Reinhoudt, D.N.; Boerrigter, H. Synthesis and Application of Iso (thio) cyanate-Functionalized Calix[4] arenes. *Liebigs Ann.* **1997**, *1997*, 2235–2245. [CrossRef]
29. Flidrova, K.; Bohm, S.; Dvorakova, H.; Eigner, V.; Lhotak, P. Dimercuration of Calix[4]arenes: Novel Substitution Pattern in Calixarene Chemistry. *Org. Lett.* **2014**, *16*, 138–141. [CrossRef] [PubMed]
30. The Binding Constants Were Calculated Using the Bindfit Application Freely. Available online: http://supramolecular.org (accessed on 30 October 2022).
31. Palatinus, L.; Chapuis, G. SUPERFLIP–a computer program for the solution of crystal structures by charge flipping in arbitrary dimensions. *J. Appl. Crystallogr.* **2007**, *40*, 786–790. [CrossRef]
32. Betteridge, P.; Carruthers, J.; Cooper, R.; Prout, K.; Watkin, D. CRYSTALS version 12: Software for guided crystal structure analysis. *J. Appl. Crystallogr.* **2003**, *36*, 1487. [CrossRef]
33. Rohlíček, J.; Hušák, M. MCE2005–a new version of a program for fast interactive visualization of electron and similar density maps optimized for small molecules. *J. Appl. Crystallogr.* **2007**, *40*, 600–601. [CrossRef]

Article

New Supramolecular Hypoxia-Sensitive Complexes Based on Azo-Thiacalixarene

Farida Galieva [1,*], Mohamed Khalifa [2,3], Zaliya Akhmetzyanova [1], Diana Mironova [2], Vladimir Burilov [2], Svetlana Solovieva [1,2,*] and Igor Antipin [1,2]

[1] Arbuzov Institute of Organic and Physical Chemistry, FRC Kazan Scientific Center, Russian Academy of Sciences, 420088 Kazan, Russia
[2] Department of Organic and Medical Chemistry, Kazan Federal University, 420008 Kazan, Russia
[3] Chemistry Department, Faculty of Science, Damanhour University, Damanhur 22511, Egypt
* Correspondence: kleo-w@mail.ru (F.G.); evgersol@yandex.ru (S.S.)

Abstract: Hypoxia accompanies many human diseases and is an indicator of tumor aggressiveness. Therefore, measuring hypoxia in vivo is clinically important. Recently, complexes of calix[4]arene were identified as potent hypoxia markers. The subject of this paper is new hypoxia-sensitive host–guest complexes of thiacalix[4]arene. We report a new high-yield synthesis method for thiacalix[4]arene with four anionic carboxyl azo fragments on the upper rim (thiacalixarene **L**) and an assessment of the complexes of thiacalixarene **L** with the most widespread cationic rhodamine dyes (6G, B, and 123) sensitivity to hypoxia. Moreover, 1D and 2D NMR spectroscopy data support the ability of the macrocycles to form complexes with dyes. Rhodamines B and 123 formed host–guest complexes of 1:1 stoichiometry. Complexes of mixed composition were formed with rhodamine 6G. The association constant between thiacalixarene **L** and rhodamine 6G is higher than for other dyes. Thiacalixarene **L**-dye complexes with rhodamine 6G and rhodamine B are stable in the presence of various substances present in a biological environment. The UV-VIS spectrometry and fluorescence showed hypoxia responsiveness of the complexes. Our results demonstrate that thiacalixarene **L** has a stronger binding with dyes compared with the previously reported azo-calix[4]arene carboxylic derivative. Thus, these results suggest higher selective visualization of hypoxia for the complexes with thiacalixarene **L**.

Keywords: supramolecular chemistry; thiacalixarene; hypoxia sensing; host–guest complex; cationic rhodamine dyes

Citation: Galieva, F.; Khalifa, M.; Akhmetzyanova, Z.; Mironova, D.; Burilov, V.; Solovieva, S.; Antipin, I. New Supramolecular Hypoxia-Sensitive Complexes Based on Azo-Thiacalixarene. *Molecules* **2023**, *28*, 466. https://doi.org/10.3390/molecules28020466

Academic Editor: Paula M. Marcos

Received: 30 November 2022
Revised: 27 December 2022
Accepted: 29 December 2022
Published: 4 January 2023

1. Introduction

According to the Global Cancer Statistics 2020, there were approximately 19.3 million new cancer cases and almost 10 million cancer deaths worldwide in 2020 [1]. Early diagnosis and treatment of cancer are crucial to reducing global mortality. A common feature of most tumors is low levels of oxygen–hypoxia [2], the severity of which depends on the type of cancer [3]. The need for oxygen exceeds the oxygen supply in the intensively proliferating and growing tumor tissue by increasing the distance in the tissue from the nearest vessel, which prevents oxygen diffusion [4]. Since the discovery of hypoxic regions in tumors in the 1950s, hypoxia has become known as a major hallmark of cancer cells and their microenvironment [5]. Hypoxia facilitates tumor growth, resistance to chemotherapy, and metastasis [6,7]. Areas with oxygen gradients and acute hypoxia appear in the vascular system and blood flow in tumor development, which induces resistance to many anticancer treatments, especially chemotherapy, radiation therapy, and photodynamic therapy [8]. Hypoxic tissues differ from normoxic in physical, chemical, and biological characteristics, such as low pH, increased expression of hypoxia-inducible factor-1α (HIF-1α), etc. [9–11].

Moreover, one of these features is an increase in the activity of oxidoreductases: nitroreductase, azoreductase, and quinone oxidoreductase [12]. Most systems proposed in the literature for the detection of hypoxia are associated with nitroaromatic and quinone derivatives [13,14]. The approach using the reduction of azo groups by azoreductase has proven itself as a promising system for visualizing hypoxia [15]. Sodium dithionite (SDT) is commonly used as the azoreductase mimic to evaluate the hypoxia-induced reduction and cleavage of the azo group since SDT can effectively and reductively cleave to the azo group [16,17]. However, most published data concern the creation of sensors utilizing a covalent binding of a dye and an azo derivative. Despite their effectiveness in creating hypoxia-sensitive systems, such sensors have some disadvantages associated with a complex molecular design, multistage, and expensive synthesis. There is also a risk of increasing the toxicity of the covalently bound compound due to the presence of a linker when the dye is introduced, as well as the formation of toxic products during the reduction of the azo group [18–20].

In this regard, the recently appeared supramolecular approach is of particular interest [21]. It consists of developing the systems based on the host–guest interaction between host azo derivatives and guest dyes (Scheme 1). The dye preliminarily interacts with azo derivatives of the macrocycles, forming a host–guest complex. Complex formation leads to quenching of the dye fluorescence. In hypoxic conditions, the azo groups of the macrocycle are reduced to amino groups, which leads to the release of four anionic fragments on the upper rim of the macrocycle and, as a result, the destruction of the host–guest complex and the restoration of dye luminescence. Given that the resulting amines are easily protonated in a wide pH range, they are not able to retain a positively charged dye, which has been shown in a number of works [22,23].

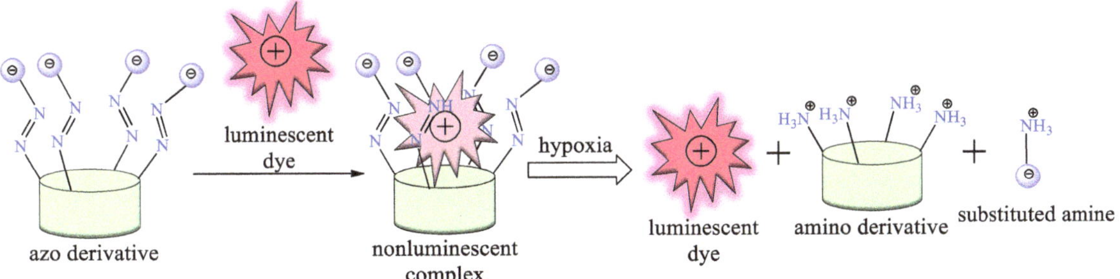

Scheme 1. The supramolecular strategy for hypoxia-sensitive systems.

The use of calix[4]arenes as a synthetic platform for the supramolecular approach seems promising due to calixarenes offer several advantages over traditional ligands: the ability to incorporate small organic molecules or a portion of large biomolecules into molecular cavities to form guest–host complexes [24]; the possibility of modification with several receptor groups and fixing their specific spatial arrangement [25,26]; various possibilities in their functionalization both at the upper and lower rims by receptor groups of different nature [27]; the absence of toxicity of both the calixarene platform and a number of derivatives of the latter [28–32]. At the same time, a few examples of the use of macrocycles to create structures capable of detecting and visualizing hypoxia are represented in the literature [33].

Recently, our research group proposed supramolecular systems for detecting hypoxia based on calix[4]arenes **1, 2** functionalized with azo groups on the upper rim (Figure 1) [34]. The present work is devoted to new hypoxia-sensitive systems using the thiacalix[4]arene platform and various rhodamine dyes. The association constant, the stoichiometry of host–guest complexes, and their absorption-emission properties under normoxic and hypoxic conditions are discussed.

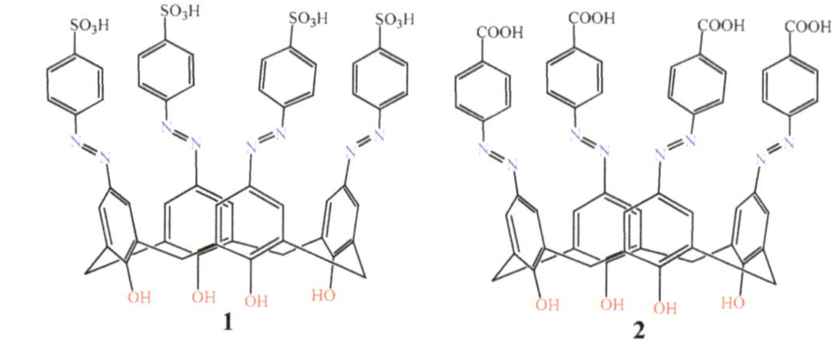

Figure 1. Structures of calix[4]arenes **1, 2** functionalized with azo groups on the upper rims [34].

2. Results

Scheme 2 shows the synthesis of azo-thiacalix[4]arene derivative **L** containing azo benzoic acid fragments. Compound **L** has been synthesized in reference [35]. However, our two-stage synthesis method used in this work provided higher yields of the target product **L**.

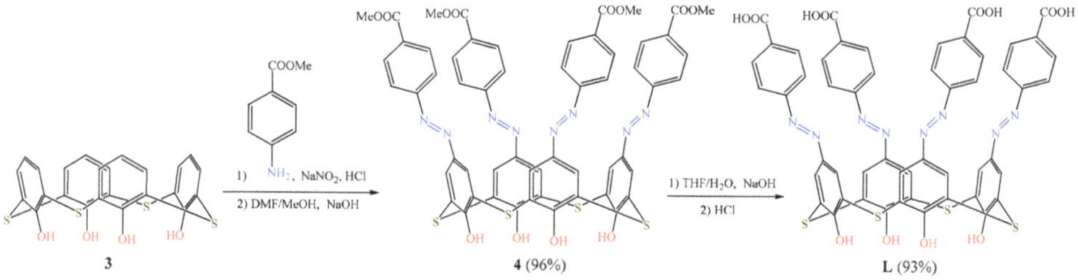

Scheme 2. Synthetic pathway for thiacalixarene **L**.

Since the macrocycle **L** contains negatively charged carboxyl groups, the most common cationic dyes, rhodamines 6G, B, and 123, were used for complexation (**Rh6G, RhB,** and **Rh123**, Figure 2). The selected dyes are characterized by high luminescence intensity and photostability [36].

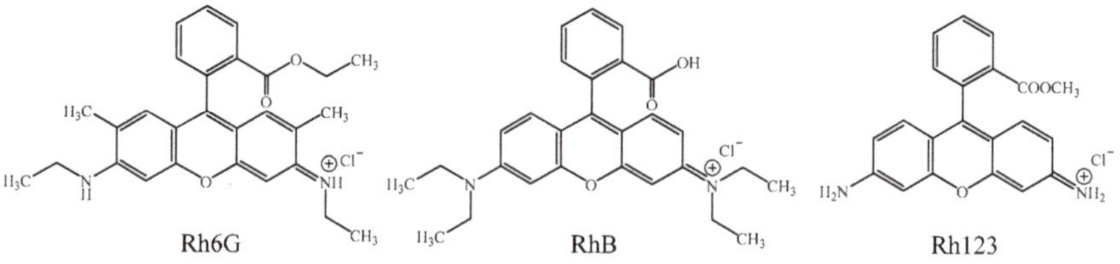

Figure 2. The structures of rhodamine dyes used in binary thiacalix[4]arene-dye systems.

The fluorescent response for rhodamine dyes upon the addition of the macrocycle **L** was determined according to fluorescence spectroscopy data. In all macrocycle–dye binary systems, quenching of dye emission was observed with an increase in the macrocycle concentration, which indicates complex formation (Figure 3). In the complex with rhodamine 6G, an increase in the concentration of the macrocycle leads to a slight hypsochromic shift of the dye maximum emission band of about 3 nm. No such effect was observed for **RhB** and **Rh123** (Figure S1).

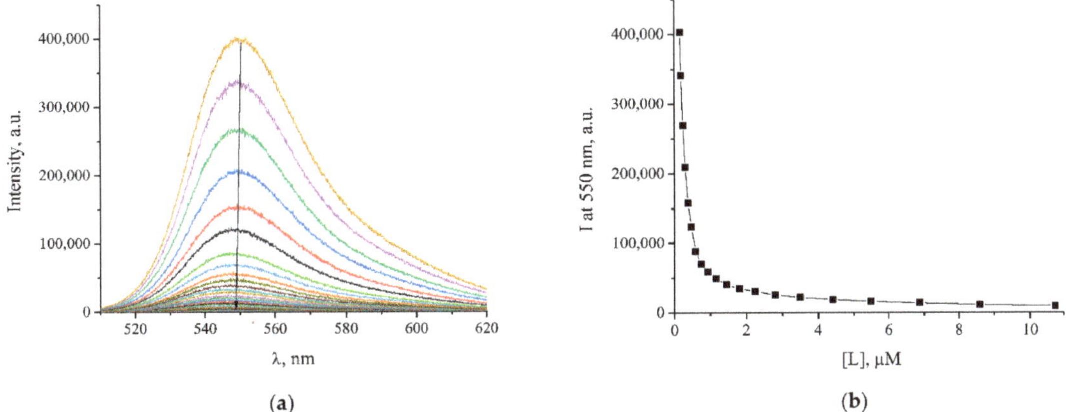

Figure 3. Direct fluorescence titration of **Rh6G** (1.0 μM) with **L** (up to 10.7 μM) (**a**) in PBS buffer (pH 7.4) at 37 °C and the associated titration curve (**b**) at λ_{em} = 550 nm.

There are two main types of fluorescence quenching mechanisms in the dye-macrocycle complex–Förster resonance energy transfer (FRET) or photoinduced electron transfer (PET). Previously, our research group showed that for calix[4]arenes **1**, **2** (Figure 1), the PET quenching mechanism prevailed [34].

Thiacalixarene **L** showed a similar trend. The linear dependence of I_0/I (I_0–the emission intensity of the free dye, I–the emission intensity of the bound dye) on the concentration of the macrocycle (Figure S2) indicates the predominance of only one type of quenching mechanism. The FRET mechanism is excluded due to the absence of overlap in the absorption band of the macrocycle with the emission band of the dye (Figure S3). Therefore, PET is the most likely quenching mechanism. To find the stoichiometry of the dye-macrocycle complexes, Job's plots were built (Figure S4). The binding stoichiometry for complexes **L** with **RhB** and **Rh123** was 1:1. In the complex of **L** with the **Rh6G** indeterminate composition, an increase in the dye proportion was found. In the complexes of calix[4]arenes **1** and **2** [34], the stoichiometry of complexes with all dyes was found as 1:1. Apparently, the increase in the number of **Rh6G** molecules (coordinated along the upper rim of the macrocycle) was affected by an increase in the size of thiacalix[4]arene cavity. It is obvious that electrostatic interactions are the driving force for the formation of such complexes. However, the preferred interaction of macrocycle **L** with **Rh6G** cannot be explained only from the standpoint of electrostatic interactions. Taking into account the great difference in the logP values of **Rh6G** and **Rh123** (6.5 and 1.5), as well as the difference in logD values (2.1. and 0.5., respectively [37], it can be assumed that the comparatively high hydrophobicity of **Rh6G** favors the interactions with the hydrophobic basket of calixarene. The association constants (K_{ass}) of azo-thiacalixarene **L** and rhodamine dyes were determined by fitting the data of the fluorescence titration (Table 1).

In the carboxylate calixarene series (**L** and **2**) the higher values of association constants with **Rh6G** were found for thiacalixarene **L**, but in absolute units, the highest association constant was observed in the case of sulfonate derivative **1** with **Rh6G**. This is associated with the electric charge distribution and the structure of the sulfate anion [38–40], providing the best electrostatic interactions with cationic dyes compared to carboxylate.

Table 1. Association constant (K_{ass}) of thiacalix[4]arene **L** and calix[4]arene **1,2** complexes with rhodamine dyes. C(dye) 1 µM, (macrocycle) 0-50 µM, PBS buffer (pH 7.4), 37 °C; [1] [34].

System	K_{ass}, M^{-1}
L-Rh6G	$4.7 \pm 0.1 \times 10^6$
L-RhB	$8.6 \pm 0.2 \times 10^4$
L-Rh123	$5.2 \pm 0.4 \times 10^3$
1-Rh6G [1]	$1.5 \pm 0.1 \times 10^7$
1-RhB [1]	$7.0 \pm 0.1 \times 10^5$
1-Rh123 [1]	$5.0 \pm 0.3 \times 10^5$
2-Rh6G [1]	$1.0 \pm 0.4 \times 10^5$
2-RhB [1]	$2.6 \pm 0.1 \times 10^4$
2-Rh123 [1]	$2.3 \pm 0.6 \times 10^4$

A comparison of absorption spectra for complexes with different dyes showed that **Rh6G** in the presence of **L** had a pronounced hypochromic effect and a 5 nm bathochromic shift of the absorption band of the dye. At the same time, **RhB** in binary **L**–dye complex had only a slight hypochromic shift whereas **Rh123** in the presence of a macrocycle practically had no changes in the absorption spectrum (Figure S5).

The ability of macrocycle **L** to form complexes with **Rh6G** was proved by ^{1}H NMR spectroscopy. In the presence of **L**, **Rh6G** protons shifts up-field (Figure 4), which is associated with shielding by the macrocycle cavity and indicates the formation of a host–guest complex with partial inclusion of the dye into the macrocyclic core.

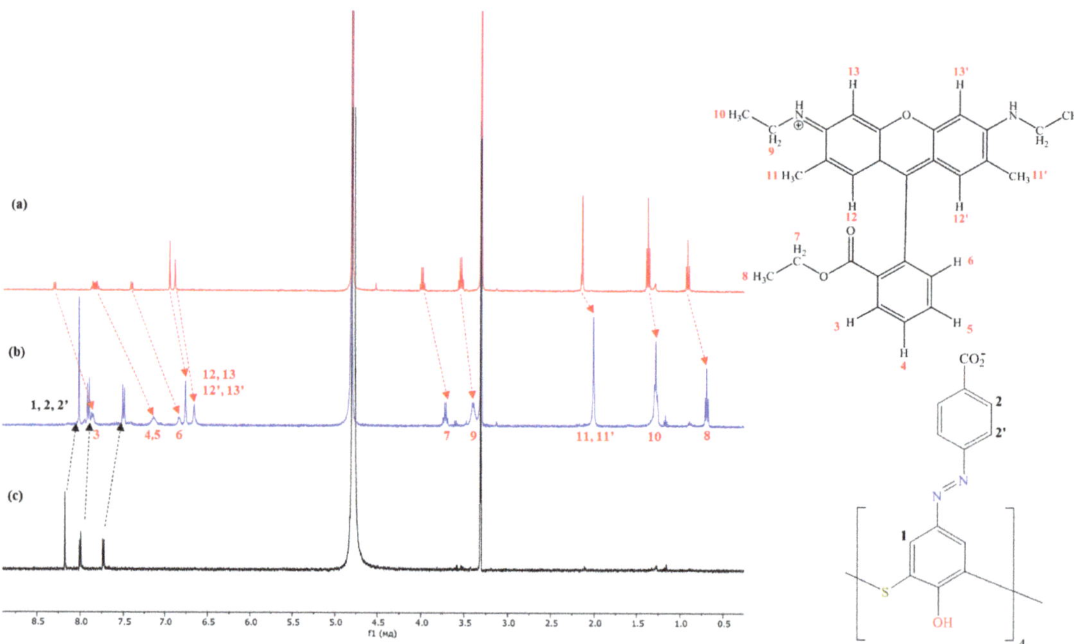

Figure 4. ^{1}H NMR spectra of 1.25 mM **Rh6G** (a), 2.5 mM thiacalixarene **L** (c), and their (0.5:1) mixture (b) in CD$_3$OD-d$_4$ at 25 °C.

The largest shifts were found for the protons of the ester-containing aromatic fragment of the **Rh6G** dye (protons 3, 4, 5, and 6 have $\Delta\delta$ = 0.44, 0.71, and 0.27 ppm, respectively) as well as for protons of the ethoxy group (7 and 8 have $\Delta\delta$ = 0.28 and 0.30 ppm, respectively). The protons of the aminoethyl and the xanthene fragments also undergo an upfield shift (protons 9, 10, 12, 13, and 11 have $\Delta\delta$ = 0.15, 0.20, 0.19, 0.32, and 0.09 ppm, respectively). Thus, it becomes apparent that the dye is immersed in the cavity of the macrocycle, forming a guest-host complex. It should be noted that the up-field shifts of dye proton signals are smaller than those for sulfonate calix[4]arene **1** [34]. However, the thiacalixarene **L** complex demonstrated much larger up-field shifts of dye proton signals compared to the complex of carboxy calix[4]arene **2-Rh123** system previously studied by the Go group [33]. Thus, the macrocycle **L** has a stronger binding than calix[4]arene **2** even though they are both bearing negatively charged carboxyl groups. This may be the result of the larger ring size of thiacalix[4]arene, which is 15% larger than that of classical calix[4]arene. Therefore thiacalix[4]arene scaffold is more flexible and easily undergoes conformational changes during the complexation. Due to the high flexibility of the molecular structure thiacalixarenes are superior to "classical" calixarenes for the formation of host–guest complexes [41]. This is consistent with the stability constants of the complexes on the thiacalix[4]arene platform and indicates a stronger interaction of carboxylate thiacalixarene **L** with rhodamine dyes. The interaction of macrocycle **L** with **Rh6G** was also confirmed by the 2D NOESY ^{1}H-^{1}H spectroscopy (Figure S6). The presence of cross peaks between aromatic protons 1 thiacalixarene **L** (δ = 7.96 ppm) with the **Rh6G** protons 3 (δ = 7.79 ppm), 6 (δ = 6.79 ppm) as well as with of protons the xanthate fragment 12-12' (δ = 6.72 ppm), 11 (δ = 1.98 ppm), protons of the aminoethyl fragment 10 (δ = 1.27 ppm) unequivocally indicates the formation of the host–guest complex. In addition, cross peaks are observed between protons 2, 2' of thiacalixarene **L** and protons 3, 6, 11-11', 13-13', and 12-12' (δ = 7.50 with 1.98, 6.60, and 6.72 ppm, respectively). This is consistent with the up-field proton shifts in ^{1}H NMR spectra (Figure 4). Thus, it can be concluded that in the complex with thiacalixarene **L**, the **Rh6G** dye is immersed in the cavity mainly by the xanthic part, while in the case of calixarene **1** the same xanthic part remains outside in the complex [34].

The strong complexation in the macrocycle–dye system is a positive feature as it reduces the undesirable slight dissociation of the carrier probe when used in difficult biological conditions. Therefore, all binary macrocycle–dye systems were tested for the presence of a fluorescent response toward various biomolecules and metal ions (Figure S7).

The intensity of dye emission does not change when the third component is added to binary systems with high values of K_a. It indicates the absence of competitive interaction between the introduced analyte and the dye in the dye–macrocycle system. Thus, systems based on **Rh6G** and **RhB**, which have no competition with bioanalytes and metal ions, are the most promising. For systems with **Rh123**, an increase in emission (up to 100%) is observed due to the displacement of dye from the macrocycle cavity by bioanalytes: lysine, ATP, BSA, and some metals (Na^+, Ca^{2+}).

UV-visible spectroscopy was used to investigate the hypoxia response of the macrocycle **L**. The thiacalixarene **L** has a broad absorption peak in the range of 300–500 nm, associated with the n-π^* transitions of the azo groups (Figure S3). The addition of an excess of sodium dithionite (SDT), which acts as a chemical imitator of azoreductase [16,17], to the macrocycle solution causes the disappearance of azo absorption, which indicates the occurrence of a reduction reaction. However, visual discoloration was not observed. Reduction kinetics were quantified by monitoring the absorbance at 370 nm in real time under normoxic (20% O_2) and hypoxic (<0.1% O_2) conditions (Figures 5 and S8).

The intensity decay curves were successfully described by the quasi-first-order reaction decay model. Table 2 presents the calculated values of the reduction constants of the azo group. For thiacalix[4]arene **L** in normal conditions, the reduction constant was higher than for calix[4]arenes **1** and **2**. It is noteworthy that hypoxic conditions lead to an increase in the azo group's reduction rate.

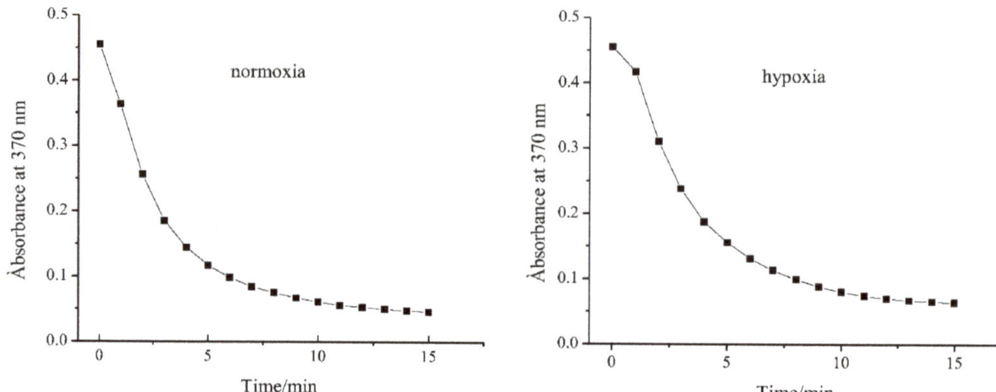

Figure 5. Absorbance at 370 nm of macrocycle **L** (10 μM) as a function of time following the addition of SDT (1.0 mM) under normoxic and hypoxic conditions, PBS buffer, pH 7.4, 37 °C.

Table 2. Reduction kinetics of the macrocycles **L**, **1**, and **2** in the presence of SDT; [1] [34].

Macrocycle	k, min^{-1} (Normoxia)	k, min^{-1} (Hypoxia)
L	0.420 (Adj. R^2 > 0.987)	0.497 (Adj. R^2 > 0.985)
1 [1]	0.350 (Adj. R^2 > 0.991)	-
2 [1]	0.403 (Adj. R^2 > 0.987)	-

In passing to binary macrocycle–dye systems, it is anticipated the interaction of the rhodamine dye itself with SDT would make a significant contribution (as shown earlier [34]), leading to the quenching of the luminescence (Figure 6).

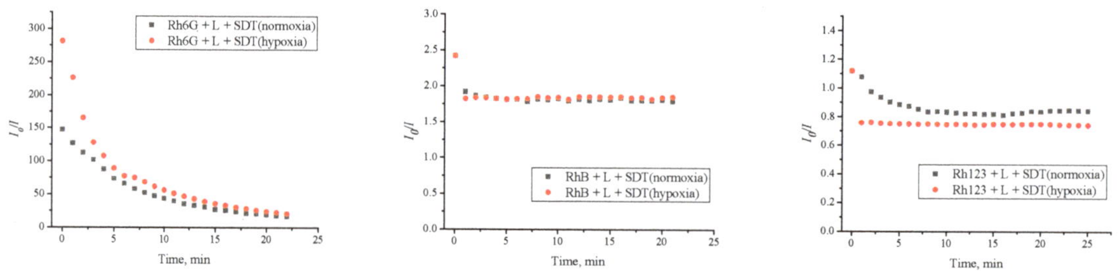

Figure 6. I_0/I vs time for rhodamine dye-thiacalixarene **L** systems in the presence of SDT under normoxia and hypoxia. I_0 is the fluorescence intensity of the free dye. C (dye) = 1 μM, C (**L**) = 1 μM, C (SDT) = 100 μM, PBS buffer (pH 7.4).

For the **RhB**–macrocycle **L** complex, the luminescence was restored to 50% (Figure S9b) of the intensity of the free dye during the first minutes, after which the luminescence intensity of the system remained unchanged for 20 min. At the same time, the results were the same in both normoxia and hypoxia. For the rhodamine-123-macrocycle complex, an increase in luminescence up to 120% of the intensity of the free dye was observed at the 8th minute in normoxic conditions and after a minute in hypoxic (Figure S9c). However, taking into account the weak complexes formation of thiacalixarene **L** with **RhB** and **Rh123**, the increase in the luminescence intensity of the systems was insignificant (Figure 6). In the complex **L** with **Rh6G**, an increase in the luminescence intensity of the system was observed within 20 min. However, the luminescence intensity reached only 5%

of the intensity for the free dye in both normoxia and hypoxia (Figure S9a). At the same time, the luminescence quenching of the complex **L–Rh6G** is more pronounced before the addition of sodium dithionite in hypoxia. The incomplete restoration of luminescence is apparently due to **Rh6G** interacting with SDT and partially transforming into the leuco form [34]. **Rh6G** luminescence quenching by SDT in the **L–Rh6G** system is less pronounced compared with calixarenes **1,2–Rh6G** complex [34]. Given the fact that even an incomplete restoration of luminescence is an 8-fold (in normoxia) and a 14-fold (in hypoxia) increase in intensity compared with the non-luminescent complex, the thiacalixarene **L-Rh6G** complex potentially can be used as a signaling system for hypoxia.

To test the specificity of the response to hypoxia, fluorescence recovery was investigated by replacing SDT with other cellular reductants (cysteine, glutathione, and reduced nicotinamide adenine dinucleotide phosphate) both in hypoxic and normoxic conditions (Figure 7). Obtained emission spectra demonstrate no fluorescent changes in the presence of an excess of selected cellular reductants. The comparison revealed that the binary macrocycle **L–Rh6G** system specifically reacts to SDT, which makes it possible to classify this system as a candidate for a hypoxia-sensitive sensor.

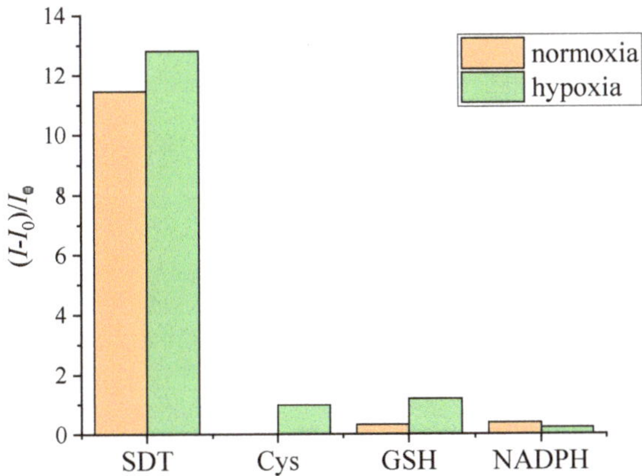

Figure 7. Fluorescent responses of **Rh6G-L** (30/10 μM) at 545 nm (λex = 490 nm) to various reductants. The complex was treated with either SDT or various biological reductants (Cys: Cysteine, GSH: glutathione, NADPH: reduced nicotinamide adenine dinucleotide phosphate) under normoxic or hypoxic conditions. Incubation time is 25 min. C (**Rh6G**) = 10 μM, C (**L**) = 30 μM, C (reductants) = 100 μM, PBS buffer (pH 7.4) at 37 °C.

3. Materials and Methods

3.1. Sample Preparation

The phosphate-buffered saline (PBS) solution with pH 7.4 was prepared by dissolving powder (Sigma P-3813) in 1000 mL of Milli-Q water. The stock solutions (100 μM) of macrocycles and dyes were prepared by dissolving the corresponding chemicals in PBS buffer. The fluorescence titrations were performed by successive addition of known amounts of macrocycle to the dye solution in a quartz cuvette. Argon gas was bubbled into the solution for 30 min to create a hypoxic environment.

3.2. Apparatus

The NMR spectra were recorded on Bruker Avance 400, 500, and 600 Nanobay (Bruker Corporation, Billerica, MA, USA) with signals from residual protons of CD_3OD-d_4 or DMSO-d_6 solvent as the internal standard. The UV-Vis spectra were recorded in a quartz cell (light path 10 mm) on a Shimadzu UV-2600 UV-Vis spectrophotometer (Shimadzu Corporation, Kyoto, Japan) equipped with a Cary dual-cell Peltier accessory. The fluorescence measurements were recorded in a conventional quartz cell (light path 10 mm) on a Fluorolog FL-221 spectrofluorometer (Horiba, Ltd., Kyoto, Japan) equipped with a single-cell Peltier accessory. MALDI mass spectra were obtained using an UltraFlex III TOF/ TOF spectrometer in the linear mode; *p*-nitroaniline was used as the matrix. Elemental analysis was performed using a PerkinElmer PE 2400 CHNS/O analyzer.

3.3. Synthesis

All reagents and solvents were purchased from either Acros (Geel, Belgium) or Sigma-Aldrich (Burlington, VT, USA) and used without further purification. The 5,11,17,23-tetrakis[(4-carboxyphenyl)azo]-25,26,27,28-tetrahydrothiacalix[4]arene **L** [35] was synthesized according to the reported methods. 5,11,17,23-tetrakis-[(4-methoxycarbonyl)-phenylazo]-25,26,27,28-tetrahydroxy-2,8,14,20-tetrathiacalix[4]arene **4**.

NaOH (25 mmol) was added to a solution of 0.248 g (0.5 mmol) of thiacalix[4]arene **3** in 30 mL of a mixture of MeOH and DMF (5:8) with constant stirring in an ice bath. The diazonium salt was obtained by diazotization of aniline benzoate (3 mmol) in 10 mL of a mixture of acetic and hydrochloric acids (1:1), water (5 mL), and $NaNO_2$ (3 mmol). The diazonium salt solution was added dropwise. The reaction mixture was stirred for 3 h at 0–5 °C, 10 h at room temperature, and 3 h when heated to 45 °C. The reaction product was precipitated from the reaction mixture by adding a solution of HCl (1:4), washed with water and methanol, and dried at room temperature for 48 h. Yield 96%. ^1H NMR (600 MHz, DMSO-d_6, 25 °C) δH ppm: 3.86 (12H, s, –C(O)O-CH$_3$), 7.87 (8H, d, J = 8.2 Hz, Ar$_{azo}$-H), 8.07 (8H, d, J = 8.2 Hz, Ar$_{azo}$-H), 8.17 (8H, s, Ar-H). MALDI TOF (m/z): 1168.0 [M + H]$^+$. Found, %: C 55.47; H 4.49; N 10.31. $C_{56}H_{40}N_8O_{12}S_4 \cdot 3H_2O \cdot 2(CH_3)_2NC(O)H$. Calculated, %: C 55.35; H 4.50; N 10.41.

4. Conclusions

In summary, we demonstrated the binary system based on azo-thiacalix[4]arene as a potential hypoxia sensor. A systematic study of the thiacalixarene–rhodamine binary system was performed by using UV-visible and fluorescence spectroscopy as well as NMR techniques. It was established that the binding stoichiometry for complexes with rhodamines B and 123 was 1:1. In the complex with rhodamine 6G, mixed complexes of indeterminate composition are formed. The association constants of thiacalixarene-rhodamine 6G complexes were higher than for other dyes. The stability of macrocycle–dye complexes with rhodamine 6G and rhodamine B was verified with the addition of analytes present in a biological environment. It was shown that the azo groups of the macrocycle are selectively reduced by sodium dithionite, which leads to the release of the dye and the restoration of its fluorescence intensity. In comparison with the carboxylate analog based on the classical calix[4]arene, the azo-thiacalix[4]arene has a stronger binding with rhodamine dyes. We expect that these results could pave the way for highly selective visualization of hypoxia in living systems.

Supplementary Materials: The following supporting information can be downloaded at: https://www.mdpi.com/article/10.3390/molecules28020466/s1, Figure S1. Direct fluorescence titration of Rh123 (a) and RhB (b) with **L** (up to 10.7 μM) in PBS buffer (pH 7.4) at 37 °C; Figure S2. Volmer plots for Rh6G, RhB, and Rh123 with thiacalix[4]arene **L**. C (dye) = 1 μM, PBS buffer (pH 7.4) at 37 °C; Figure S3. Normalized emission spectrum of Rho6G and absorption spectrum of thiacalix[4]arene **L** in PBS buffer (pH 7.4) at 37 °C; Figure S4. Job's plot for solutions rhodamine dyes and thiacalix[4]arene **L**. C [total] = 2 μM, PBS buffer (pH 7.4) at 37 °C; Figure S5. Absorbance spectra of dyes and binary thiacalix[4]arene **L**-dye systems. C [dye] = [calixarene] = 10 μM, PBS buffer (pH 7.4) at 37 °C; Figure S6. A fragment of 2D NOESY ^{1}H-^{1}H spectra of thiacalix[4]arene **L** with Rh6G, 1.25 mM Rho6G and 2.5 mM of thiacalix[4]arene **L** in CD$_3$OD-d$_4$ at 25 °C; Figure S7. Fluorescent response of binary dye–thiacalix[4]arene **L** systems upon the addition of various biologically coexisting species. I_0 and I are the fluorescence intensity before and after the addition of various substances present in biological media, respectively. In the last right column, I is the fluorescence intensity of the free dyes. C [dye] = 1 μM, C [**L**] = 3 μM, C [substance] = 10 μM, PBS buffer (pH 7.4) at 37 °C. Figure S8. Absorbance spectra of **L** (10 μM) before and after reduction by SDT (1.0 mM) under normoxic (20%

O_2) (a) and hypoxic (less than 0.1% O_2) (b) conditions for 15 min, PBS buffer (pH 7.4) at 37 °C.
Figure S9. Fluorescent spectra of dye, the complex before the addition of SDT, and after 20 min of
exposure with the addition of SDT for systems with Rh6G (a), RhB (b), and Rh123 (c). C [dye] = 1 μM,
C [L] = 1 μM, C [SDT] = 100 μM, PBS buffer (pH 7.4) at 37 °C.

Author Contributions: Conceptualization, F.G., D.M., V.B.; investigation, F.G., M.K. and Z.A.; data
curation, F.G., D.M. and V.B.; writing—original draft preparation, F.G., D.M. and V.B.; writing—
review and editing, S.S. and I.A.; supervision, S.S. and I.A.; project administration, F.G. All authors
have read and agreed to the published version of the manuscript.

Funding: This research was funded by the Russian Science Foundation, grant number 22-73-00138.

Institutional Review Board Statement: Not applicable.

Informed Consent Statement: Not applicable.

Data Availability Statement: The data presented in this study are contained within the article or in
the Supplementary Materials, or are available upon request from the corresponding authors (Farida
Galieva and Svetlana Solovieva).

Acknowledgments: The authors are grateful to the Assigned Spectral-Analytical Center of Shared
Facilities for Study of Structure, Composition and Properties of Substances and Materials of the
Federal Research Center of Kazan Scientific Center of Russian Academy of Sciences (CSF-SAC FRC
KSC RAS) for the technical support.

Conflicts of Interest: The authors declare no conflict of interest.

Sample Availability: Samples of the compound **L** are available from the authors.

References

1. Sung, H.; Ferlay, J.; Siegel, R.L.; Laversanne, M.; Soerjomataram, I.; Jemal, A.; Bray, F. Global cancer statistics 2020: GLOBOCAN estimates of incidence and mortality worldwide for 36 cancers in 185 countries. *CA A Cancer J. Clin.* **2021**, *71*, 209–249. [CrossRef] [PubMed]
2. Lee, J.W.; Ko, J.; Ju, C.; Eltzschig, H.K. Hypoxia signaling in human diseases and therapeutic targets. *Exp. Mol. Med.* **2019**, *51*, 1–13. [CrossRef] [PubMed]
3. Bhandari, V.; Hoey, C.; Liu, L.Y.; Lalonde, E.; Ray, J.; Livingstone, J.; Lesurf, R.; Shiah, Y.J.; Vujcic, T.; Huang, X.Y.; et al. Molecular landmarks of tumor hypoxia across cancer types. *Nat. Genet.* **2019**, *51*, 308–318. [CrossRef] [PubMed]
4. Semenza, G.L. Hypoxia, clonal selection, and the role of HIF-1 in tumor progression. *Crit. Rev. Biochem. Mol. Biol.* **2000**, *35*, 71–103. [CrossRef] [PubMed]
5. Thomlinson, R.H.; Gray, L.H. The histological structure of some human lung cancers and the possible implications for radiotherapy. *Br. J. Cancer* **1955**, *9*, 539–549. [CrossRef] [PubMed]
6. Hayashi, Y.; Yokota, A.; Harada, H.; Huang, G. Hypoxia/pseudohypoxia-mediated activation of hypoxia-inducible factor-1α in cancer. *Cancer Sci.* **2019**, *110*, 1510–1517. [CrossRef]
7. Gilkes, D.M.; Semenza, G.L. Role of hypoxia-inducible factors in breast cancer metastasis. *Future Oncol.* **2013**, *9*, 1623–1636. [CrossRef] [PubMed]
8. Brown, J.M.; William, W.R. Exploiting tumour hypoxia in cancer treatment. *Nat. Rev. Cancer* **2004**, *4*, 437–447. [CrossRef]
9. Petrova, V.; Annicchiarico-Petruzzelli, M.; Melino, G.; Amelio, I. The hypoxic tumour microenvironment. *Oncogenesis* **2018**, *7*, 1–13. [CrossRef]
10. Avagliano, A.; Granato, G.; Ruocco, M.R.; Romano, V.; Belviso, I.; Carfora, A.; Montagnani, S.; Arcucci, A. Metabolic reprogramming of cancer associated fibroblasts: The slavery of stromal fibroblasts. *BioMed Res. Int.* **2018**, *111*, 1–12. [CrossRef]
11. Ammirante, M.; Shalapour, S.; Kang, Y.; Jamieson, C.A.M.; Karin, M. Tissue injury and hypoxia promote malignant progression of prostate cancer by inducing CXCL13 expression in tumor myofibroblasts. *Proc. Natl. Acad. Sci. USA* **2014**, *111*, 14776–14781. [CrossRef] [PubMed]
12. Zhou, H.; Qin, F.; Chen, C. Designing hypoxia-responsive nanotheranostic agents for tumor imaging and therapy. *Adv. Healthc. Mater.* **2021**, *10*, 1–28. [CrossRef] [PubMed]
13. Tanabe, K.; Hirata, N.; Harada, H.; Hiraoka, M.; Nishimoto, S.-I. Emission under hypoxia: One-electron reduction and fluorescence characteristics of an indolequinone-coumarin conjugate. *Chembiochem* **2008**, *9*, 426–432. [CrossRef] [PubMed]
14. Liu, Y.; Xu, Y.F.; Qian, X.H.; Liu, J.W.; Shen, L.Y.; Li, J.H.; Zhang, Y.X. Novel fluorescent markers for hypoxic cells of naphthalimides with two heterocyclic side chains for bioreductive binding. *Bioorg. Med. Chem.* **2006**, *14*, 2935–2941. [CrossRef]
15. Kumari, R.; Sunil, D.; Ningthoujam, R.S.; Kumar, N.V.A. Azodyes as markers for tumor hypoxia imaging and therapy: An up-to-date review. *Chem.-Biol. Interact.* **2019**, *207*, 91–104. [CrossRef]

16. Chevalier, A.; Mercier, C.; Saurel, L.; Orenga, S.; Renard, P.-Y.; Romieu, A. The first latent green fluorophores for the detection of azoreductase activity in bacterial cultures. *Chem. Commun.* **2013**, *49*, 8815. [CrossRef]

17. Leriche, G.; Budin, G.; Darwich, Z.; Weltin, D.; Mély, Y.; Klymchenko, A.S.; Wagner, A. A FRET-based probe with a chemically deactivatable quencher. *Chem. Commun.* **2012**, *48*, 3224–3226. [CrossRef]

18. Verwilst, P.; Han, J.; Lee, J.; Mun, S.; Kang, H.-G.; Kim, J.S. Reconsidering azobenzene as a component of small-molecule hypoxia-mediated cancer drugs: A theranostic case study. *Biomaterials* **2017**, *115*, 104–114. [CrossRef]

19. Yuan, P.; Zhang, H.; Qian, L.; Mao, X.; Du, S.; Yu, C.; Peng, B.; Yao, S.Q. Intracellular delivery of functional native antibodies under hypoxic conditions by using a biodegradable silica nanoquencher. *Angew. Chem. Int. Ed.* **2017**, *56*, 12481–12485. [CrossRef]

20. Perche, F.; Biswas, S.; Wang, T.; Zhu, L.; Torchilin, V.P. Hypoxia-targeted siRNA delivery. *Angew. Chem. Int. Ed.* **2014**, *53*, 3362–3366. [CrossRef]

21. Ma, X.; Zhao, Y.L. Biomedical applications of supramolecular systems based on host-guest interactions. *Chem. Rev.* **2015**, *115*, 7794–7839. [CrossRef]

22. Chevalier, A.; Piao, W.; Hanaoka, K.; Nagano, T.; Renard, P.-Y.; Romieu, A. Azobenzene-caged sulforhodamine dyes: A novel class of 'turn-on' reactive probes for hypoxic tumor cell imaging. *Methods Appl. Fluores* **2015**, *3*, 044004. [CrossRef]

23. Piao, W.; Tsuda, S.; Tanaka, Y.; Maeda, S.; Liu, F.; Takahashi, S.; Kushida, Y.; Komatsu, T.; Ueno, T.; Terai, T.; et al. Development of Azo-Based Fluorescent Probes to Detect Different Levels of Hypoxia. *Angew. Chem. Int. Ed.* **2013**, *52*, 13028–13032. [CrossRef]

24. Tauran, Y.; Coleman, A.W.; Perret, F.; Kim, B. Cellular and in vivo biological activities of the calix[n]arenes. *Curr. Org. Chem.* **2015**, *19*, 2250–2270. [CrossRef]

25. Nimse, S.B.; Kim, T. Biological applications of functionalized calixarenes. *Chem. Soc. Rev.* **2013**, *42*, 366–386. [CrossRef] [PubMed]

26. Baldini, L.; Casnati, A.; Sansone, F. Multivalent and multifunctional calixarenes in bionanotechnology. *Eur. J. Org. Chem.* **2020**, *2020*, 5056–5069. [CrossRef]

27. Tian, H.W.; Liu, Y.C.; Guo, D.S. Assembling features of calixarene-based amphiphiles and supra-amphiphiles. *Mater. Chem. Front.* **2020**, *4*, 46–98. [CrossRef]

28. Mokhtari, B.; Pourabdollah, K. Applications of nano-baskets in drug development: High solubility and low toxicity. *Drug Chem. Toxicol.* **2013**, *36*, 119–132. [CrossRef]

29. Ukhatskaya, E.V.; Kurkov, S.V.; Matthews, S.E.; Loftsson, T. Encapsulation of drug molecules into calix[n]arene nanobaskets. Role of aminocalix[n]arenes in biopharmaceutical field. *J. Pharm. Sci.* **2013**, *102*, 3485–3512. [CrossRef]

30. Coleman, A.W.; Jebors, S.; Cecillon, S.; Perret, P.; Garin, D.; Marti-Battle, D.; Moulin, M. Toxicity and biodistribution of para-sulfonato-calix[4]arene in mice. *New J. Chem.* **2008**, *32*, 780–782. [CrossRef]

31. Perret, F.; Coleman, A.W. Biochemistry of anionic calix[n]arenes. *Chem. Commun.* **2011**, *47*, 7303–7319. [CrossRef]

32. Perret, F.; Lazar, A.N.; Coleman, A.W. Biochemistry of the para-sulfonato-calix[n]arenes. *Chem. Commun.* **2006**, *23*, 2425–2438. [CrossRef]

33. Geng, W.-C.; Jia, S.; Zheng, Z.; Li, Z.; Ding, D.; Guo, D.-S. A noncovalent fluorescence turn-on strategy for hypoxia imaging. *Angew. Chem. Int. Ed.* **2019**, *58*, 2377–2381. [CrossRef]

34. Mironova, D.; Burilov, B.; Galieva, F.; Khalifa, M.A.M.; Kleshnina, S.; Gazalieva, A.; Nugmanov, R.; Solovieva, S.; Antipin, I. Azocalix[4]arene-rhodamine supramolecular hypoxia-sensitive systems: A search for the best calixarene hosts and rhodamine guests. *Molecules* **2021**, *26*, 5451. [CrossRef]

35. Chakrabarti, A.; Chawla, H.M.; Francis, T.; Pant, N.; Upreti, S. Synthesis and cation binding properties of new arylazo- and heteroarylazotetrathiacalix[4]arenes. *Tetrahedron* **2006**, *62*, 1150–1157. [CrossRef]

36. Rajasekar, M. Recent Trends in Rhodamine derivatives as fluorescent probes for biomaterial applications. *J. Mol. Struct.* **2021**, *1235*, 130232. [CrossRef]

37. Kennedy, A.R.; Conway, L.K.; Kirkhouse, J.B.A.; McCarney, K.M.; Puissegur, O.; Staunton, E.; Teat, S.J.; Warren, J.E. Monosulfonated Azo Dyes: A Crystallographic Study of the Molecular Structures of the Free Acid, Anionic and Dianionic Forms. *Crystals* **2020**, *10*, 662. [CrossRef]

38. Zana, R.; Schmidt, J.; Talmon, Y. Tetrabutylammonium Alkyl Carboxylate Surfactants in Aqueous Solution: Self-Association Behavior, Solution Nanostructure, and Comparison with Tetrabutylammonium Alkyl Sulfate Surfactants. *Langmuir* **2005**, *21*, 11628–11636. [CrossRef] [PubMed]

39. Reshetnyak, E.A.; Chernysheva, O.S.; Nikitina, N.A.; Loginova, L.P.; Mchedlov-Petrosyan, N.O. Activity coefficients of alkyl sulfate and alkylsulfonate ions in aqueous and water-salt premicellar solutions. *Colloid J.* **2014**, *76*, 358–365. [CrossRef]

40. Burilov, V.A.; Mironova, D.A.; Ibragimova, R.R.; Nugmanov, R.I.; Solovieva, S.E.; Antipin, I.S. Detection of sulfate surface-active substances via fluorescent response using new amphiphilic thiacalix[4]arenes bearing cationic headgroups with Eosin Y dye. *Colloids Surf. A Physicochem. Eng. Asp.* **2017**, *515*, 41–49. [CrossRef]

41. Solovieva, S.E.; Burilov, V.A.; Antipin, I.S. Thiacalix[4]arene's Lower Rim Derivatives: Synthesis and Supramolecular Properties. *Macroheterocycles* **2017**, *10*, 134–146. [CrossRef]

Article

Calix[4]arene Derivative for Iodine Capture and Effect on Leaching of Iodine through Packaging

Loredana Ferreri [1,†], Marco Rapisarda [2,†], Melania Leanza [2], Cristina Munzone [3], Nicola D'Antona [1], Grazia Maria Letizia Consoli [1,*], Paola Rizzarelli [2,*] and Emanuela Teresa Agata Spina [2]

1   Institute of Biomolecular Chemistry, CNR, Via Paolo Gaifami 18, 95126 Catania, Italy
2   Institute for Polymers, Composites and Biomaterials, CNR, Via Paolo Gaifami 18, 95126 Catania, Italy
3   Department of Chemistry, University of Catania, Viale A. Doria 6, 95125 Catania, Italy
*   Correspondence: grazia.consoli@icb.cnr.it (G.M.L.C.); paola.rizzarelli@cnr.it (P.R.);
    Tel.: +39-0957338319 (G.M.L.C.); +39-0957338236 (P.R.)
†   These authors contributed equally to this work.

Abstract: A hydrophobic calix[4]arene derivative was investigated for its iodine (I_2) capture efficiency from gaseous and liquid phase. The iodine uptake was followed by UV-vis spectroscopy. Additionally, the influence of the calix[4]arene derivative–polyolefin system on the leaching of iodine through packaging from a povidone-iodine-based (PVP-I) formulation was evaluated. In fact, iodine is a low-cost, multi-target, and broad-spectrum antiseptic. However, it is volatile, and the extended storage of I_2-based formulations is challenging in plastic packaging. Here, we investigated the possibility of reducing the loss of I_2 from an iodophor formulation by incorporating 4-*tert*-butylcalix [4]arene-tetraacetic acid tetraethyl ester (CX) and its iodine complex in high-density polyethylene (HDPE) or polypropylene (PP) via a swelling procedure. Surface and bulk changes were monitored by contact angle, thermogravimetric analysis (TGA), and UV-vis diffuse reflectance spectra. The barrier effect of the different polymeric systems (embedded with CX, iodine-CX complex, or I2) was evaluated by monitoring the I2 retention in a buffered PVP-I solution by UV-vis spectroscopy. Overall, experimental data showed the capability of the calix[4]arene derivative to complex iodine in solution and the solid state and a significant reduction in the iodine leaching by the PP-CX systems.

Keywords: iodine capture; calix[4]arene; povidone-iodine; polyethylene; polypropylene; iodophore; plastic packaging

Citation: Ferreri, L.; Rapisarda, M.; Leanza, M.; Munzone, C.; D'Antona, N.; Consoli, G.M.L.; Rizzarelli, P.; Spina, E.T.A. Calix[4]arene Derivative for Iodine Capture and Effect on Leaching of Iodine through Packaging. *Molecules* **2023**, *28*, 1869. https://doi.org/10.3390/molecules28041869

Academic Editor: Paula M. Marcos

Received: 2 February 2023
Revised: 13 February 2023
Accepted: 15 February 2023
Published: 16 February 2023

1. Introduction

Calix[n]arenes are a family of macrocyclic polyphenols characterized by synthetic versatility and the presence of an aromatic cavity that can host ions and neutral molecules [1]. The *p-tert*-butyl-calix[4]arene, the smallest oligomer of this family, can be blocked in a cone conformation by functionalization of the phenolic OH groups at the lower rim of the macrocycle [2]. The rigid 'basket' structure improves the possibilities of preparing host–guest or inclusion complexes [3]. The capability of the π-electron-rich cavity of the calix[4]arene macrocycle to complex iodine by halogen-π interactions has been reported. Calix[4]arene derivatives [4,5] and polymeric calixarenes [6–8] have provided functional compounds or nanosheets adsorbing iodine in vapor and solution phase with applications for pollutant removal [9–11].

Iodine has been known as an effective bactericide since 1800. However, its widespread use was limited by several undesirable factors, such as poor water solubility, limited chemical stability, and high local toxicity. Therefore, a new class of chemical complexes of iodine with organic polymers, known as iodophors, were investigated [12]. In particular, the complex of iodine with polyvinilpirrolidone (PVP-I) is largely utilized clinically due to its broad range of antimicrobial activity and unknown bacterial resistance. Povidone-iodine complex consists of PVP units that are linked with iodine via hydrogen bonds between two pyrroles and contains triiodide anions and a small amount of non-complexed mobile iodine (I_2-free) that is the active bactericidal agent [13,14].

One of the main problems faced in the pharmaceutical formulations of PVP-I, as well as other organic iodophor solutions packaged in a plastic container for medical use, is the leaching of I_2 through the packaging itself [15]. This involves both a decrease in stability and therapeutic capacity of the iodophor solutions. This problem mainly concerns low-concentration PVP-I solutions necessary for the treatment of delicate organs such as the eyes [15,16]. Indeed, the loss of iodine is a function of the packaging material for a given temperature and PVP-I concentration. In particular, the plastic containers in low-density polyethylene (LDPE) are more permeable to iodine than high-density polyethylene (HDPE) or polypropylene (PP) [17]. Glass (not permeable to iodine) represents the more suitable material for a container to store dilute PVP-I solutions. However, not even the glass containers can totally maintain, for a prolonged period, the stability of dilute solutions because of the plastic dropper, which may be an important source for the leaching of iodine.

To the best of our knowledge, calixarene derivatives have been employed as efficient catalysts for the polymerization of propylene [18,19] or stabilizers against PP oxidative degradation [20]. They have been grafted on PP [21] or mixed with PP [22], but never embedded in PP by the swelling method. In this study, we successfully investigated the capacity of a hydrophobic calix[4]arene derivative (tetra(ethoxycarbonyl-methoxy)-4-*tert*-butylcalix[4]arene, CX) to capture iodine from gaseous and liquid phase. Moreover, we explored the possibility of reducing the leaching of I_2 through packaging from a PVP-I solution by embedding CX and its iodine complex in polymer matrices (HDPE or PP) via a swelling procedure. Comparative tests were carried out including polymer films embedded with I_2 only. Surface and bulk changes were monitored by different analytical techniques (contact angle, TGA, UV). The barrier effect of the different polymeric systems (PP embedded with CX, iodine-loaded CX or I_2) was evaluated by monitoring the iodine retention in a buffered PVP-I solution by UV-vis spectroscopy. The iodine complex of CX embedded in the PP films showed good stability over time and better reduced the leaching of iodine from the PVP-I buffered solution in comparison with unmodified polyolefin samples (barrier effect).

2. Results

2.1. Synthesis of Tetra(ethoxycarbonyl-methoxy)-4-tert-butylcalix[4]arene

tetra(ethoxycarbonyl-methoxy)-4-*tert*-butylcalix[4]arene (CX) was synthesized by following a procedure found in the literature [23]. In brief, the commercial 4-*tert*-butylcalix[4]arene was treated with bromo-ethylacetate in the presence of potassium carbonate (molar ratio 1:10:10) in acetone as a solvent. CX was characterized by ^{1}H NMR spectrum whose signals (Figure S1) were consistent with the structure depicted in Figure 1.

2.2. Formation of CX/Iodine Inclusion Complex

Iodine is one of the strongest electron acceptors. In principle, CX can host iodine in its π-electron-rich cavity and form a charge transfer complex. Accordingly, we explored the capability of CX to capture iodine in chloroform solution and in the solid state from iodine water solution (solid–liquid method) and iodine vapor (solid–air method).

2.2.1. Formation of the Complex in Chloroform Solution

To investigate the formation of a complex between CX and iodine, chloroform was chosen as an ideal solvent. The addition of molecular iodine to a chloroform solution of CX showed the iodine absorption band at 510 nm and the appearance of a new band at 366 nm (Figure 2) relative to the formation of the CX/iodine complex. This behavior was reported for other complexes of calix[4]arene derivatives with iodine [4,5]. The absorption at 366 nm was enhanced by increasing the molar excess of CX (from 1:0.5 to 1:3 molar ratio) (Figure S2).

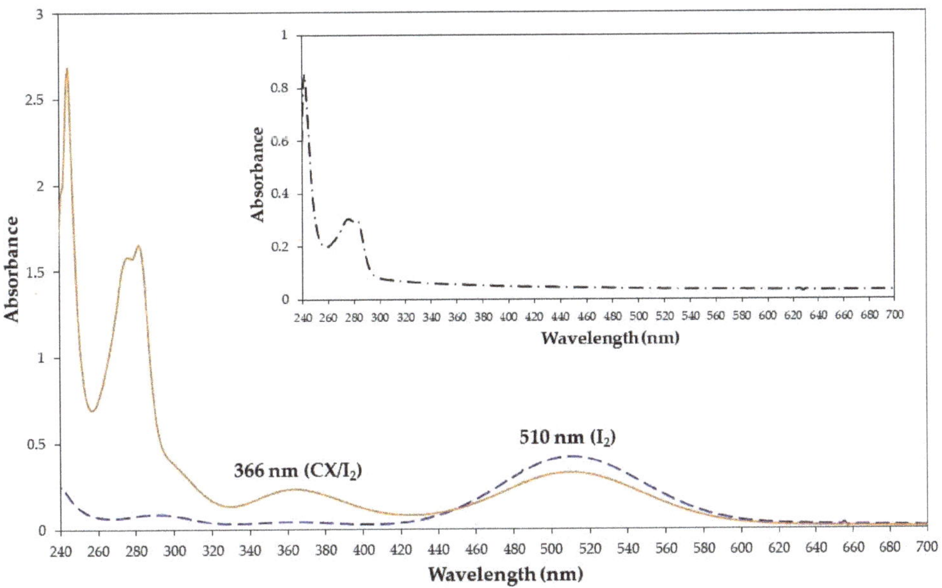

Figure 1. Molecular structure of the tetra(ethoxycarbonyl-methoxy)-4-*tert*-butylcalix[4]arene (CX).

Figure 2. UV-vis spectra of iodine (1.3×10^{-3} M, dashed blue line), CX/iodine 1:1 molar ratio (solid orange line), and CX alone (inset) in chloroform at 25 °C.

Similar to other calix[4]arene derivatives, the binding stoichiometry of the CX/iodine complex was determined to be 1:1 and a log K of 3.95 was found for the complex formation constant (see ESI and Figures S3–S5).

2.2.2. Formation of the Complex by Solid CX and Iodine in Water Solution (Solid–Liquid Method)

The water insoluble CX also showed the capability to capture iodine from an iodine water solution. When an excess (10:1 molar ratio) of CX (55 mg/mL) was added to a water solution of iodine (0.3 mg/mL), the white calixarene powder turned orange and the yellow water solution became discolored (Figure 3, inset). The capture of iodine by CX, visible by the naked eye, was corroborated by the comparison of the UV-vis spectra of the iodine water solution before and after the contact with CX (Figure 3). The absorbance of the iodine in water was reduced by 98% after the contact with CX. As further evidence of the complex formation, the UV-vis spectrum of the orange powder dissolved in chloroform showed the typical spectrum of the complex with absorption at 366 nm and no significant absorption at 510 nm relative to free iodine (Figure 3, inset).

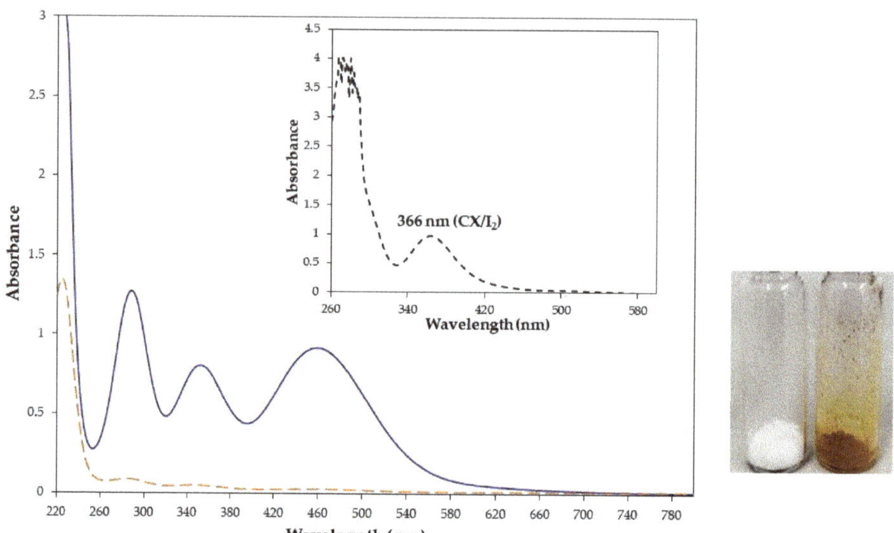

Figure 3. UV-vis spectra of iodine water solution (0.33 mg/mL, 1.3×10^{-3} M) before (solid blue line) and after stirring with CX (dashed orange line) and CX/iodine complex powder dissolved in chloroform (inset). Pictures of CX (white powder) and its complex with iodine after filtration (orange powder).

The amount of iodine captured from the water solution depends on the CX amount, as evidenced by monitoring the reduction in the iodine absorption bands in water at increasing amounts of CX (Table 1 and Figure S6).

Table 1. Reduction (%) of the absorbance of the iodine bands in water solution at different iodine:CX molar ratio.

Wavelength (nm)	Absorbance Reduction %		
	1:2 Molar Ratio	**1:6 Molar Ratio**	**1:10 Molar Ratio**
288	−36.4%	−61.0%	−93.4%
352	−35.0%	−61.0%	−94.3%
457	−66.3%	−92.5 %	−97.4%

Additionally, in agreement with the 1:1 stoichiometry of the complex, when CX was added to an excess of iodine (iodine:CX, 2:1 molar ratio) in a water solution, the capture of iodine by CX reduced the iodine absorbance by 50% (Figure S7a). As a further confirmation, a reduction by 50% was also found for the absorption of iodine extracted from the water solution by cyclohexane before and after contact with the CX powder (Figure S7b).

Similar results were obtained in phosphate/citrate buffer solution (pH 6). Decreases of 37, 73, 72, and 68% were recorded for the iodine absorption bands at 231 nm, 288 nm, 352 nm, and 453 nm, respectively, after contact with CX powder (iodine:CX, 1:2 molar ratio).

2.2.3. Formation of the Complex by Solid CX and Iodine in Vapor form (Solid–Air Method)

A magnetically stirred powder of CX (50 mg) was placed overnight at 75 °C in a closed container with solid iodine, which sublimates and produces I_2 vapor. The powder color changed from white to orange and color intensity increased over time. The UV-vis spectrum of the powder dissolved in chloroform showed the typical spectrum of the CX/iodine complex (Figure S8).

The complex formation was monitored by UV-vis analysis in chloroform at different interval times. The absorbance values at 366 nm were plotted as a function of time (Figure 4). The trend in Figure 4 shows that the complex formation proceeded up to 80 min, after which a reduction in the absorbance was observed. The UV-vis spectrum in chloroform of the sample at 100 min showed, in addition to the band at 366 nm, a band at 510 nm relative to free iodine. This behavior, also observed for other iodophors, indicated that after saturation the adsorption of I_2 is accompanied with iodine release [9].

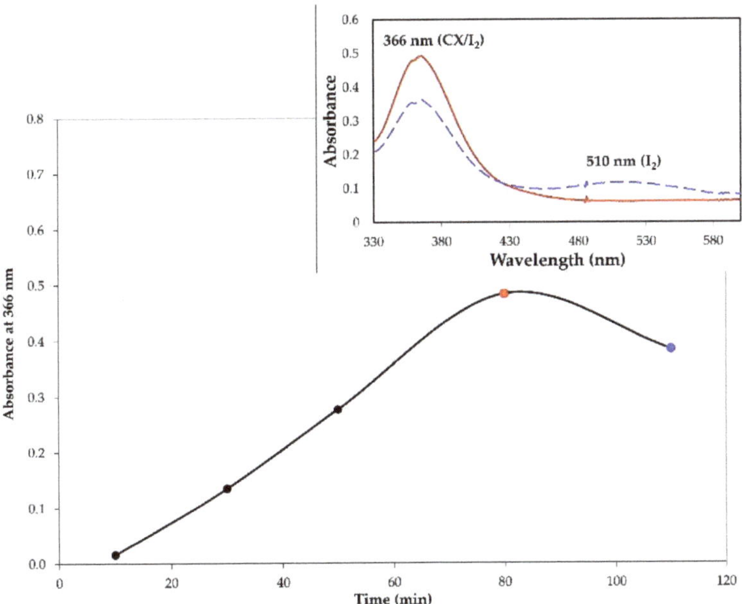

Figure 4. Monitoring of the CX/iodine complex formation by solid–air method. The graphic reports the absorbance values of the complex (λ = 366 nm) as a function of time. Inset: UV-vis spectrum in chloroform of CX/iodine complex after 80 min (solid red line) and 100 min (dashed blue line).

The complexation of I_2 by CX was corroborated by reflectance (%) spectra. Figure 5 shows the different profile of the reflectance spectra of the solid samples of CX (Figure 5a), iodine (Figure 5b), and the CX/iodine complex (Figure 5c). Typical bands centered at 238 and 274 nm and at 230, 374, 478, and 636 nm were observed for CX and iodine, respectively. The spectrum of the complex (Figure 5c) showed reflectance bands at 243 and 275 nm, ascribable to CX, and 368 and 478 nm, relative to the iodine species.

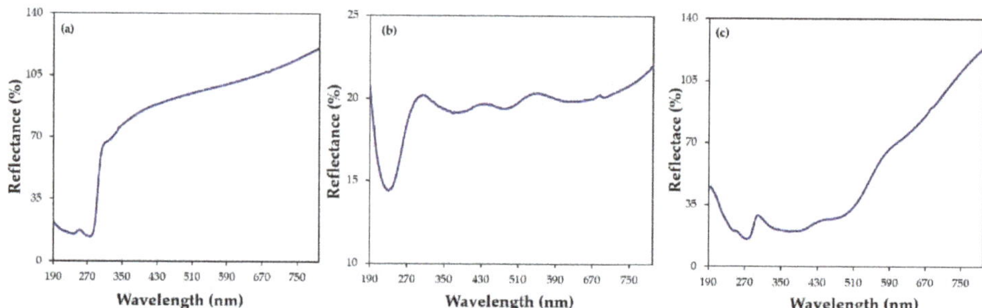

Figure 5. Reflectance (%) spectra of the (**a**) CX, (**b**) iodine, and (**c**) CX/iodine complex powders.

2.3. Stability of the CX/Iodine Inclusion Complex

The thermal stability of CX and its iodine inclusion complexes was evaluated by TGA. Figure 6 shows their weight loss percentage and derivatives as a function of the temperature. TGA of CX has two degradation steps (Figure 6a), the main being at 344 °C. TGA of both complexes from water solution (Figure 6b) and vapor (Figure 6c) highlighted an additional degradation step at about 128 °C due to iodine (Figure S9).

2.4. Calix[4]arene-Polyolefin Systems

Considering the gas permeation mechanism and assuming that the loss of I_2 is the determining factor for packaging in plastic material, we evaluated different approaches that could repress or reduce the diffusion of I_2, such as the introduction into the packaging of an additional amount of iodine. The established capacity and stability of CX to complex iodine suggested testing it in comparison with I_2.

Therefore, CX and its iodine inclusion complex were embedded in polymer matrices (HDPE or PP) via a swelling procedure. Comparative tests were carried out incorporating I_2 in HDPE or PP. Preliminary tests at room temperature and in pure solvent (Figure S10) were carried out to establish the necessary swelling time.

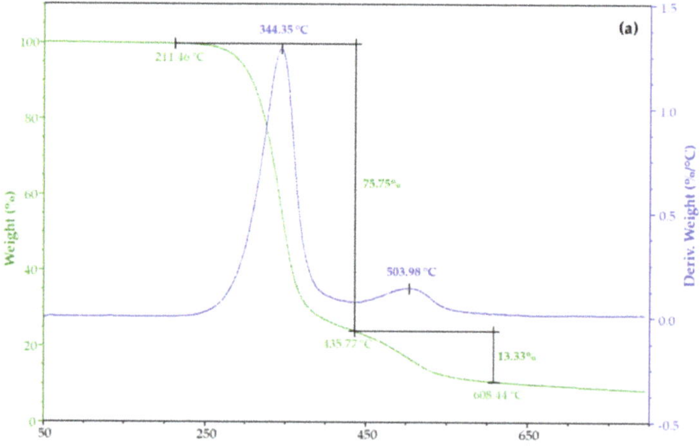

Figure 6. *Cont.*

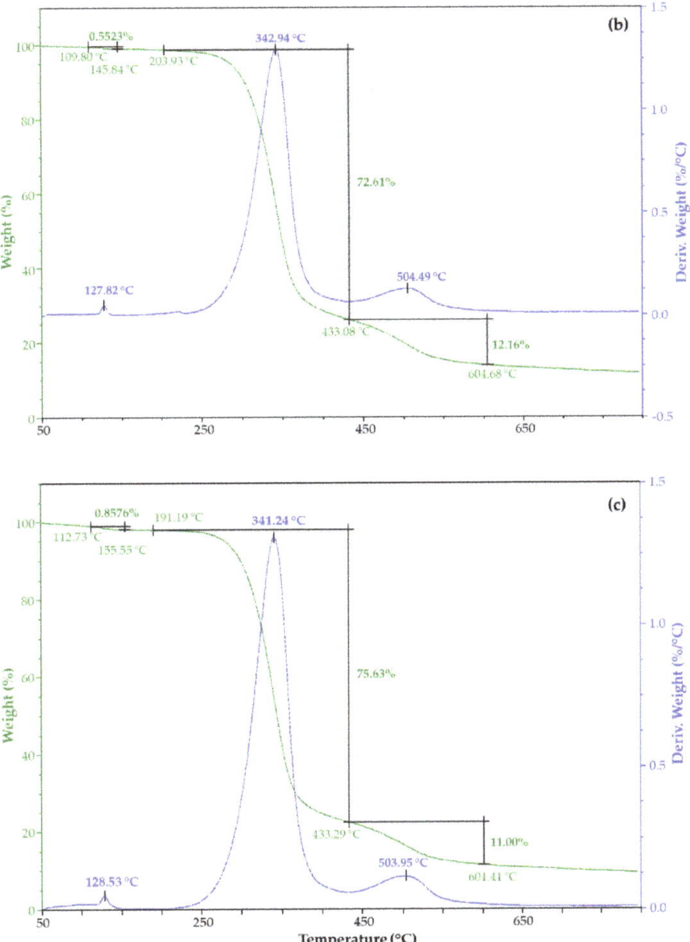

Figure 6. TGA and DTG of (**a**) CX and its complex from iodine, (**b**) water solution, and (**c**) vapor.

Consequently, polyolefin film portions were dipped in chloroform solutions of CX, CX/iodine complex, or I_2 for 24 h at room temperature. After solvent evaporation, surface and bulk changes in the film portions were monitored by contact angle, TGA, and UV-vis.

The embedding of CX in HDPE or PP was performed by dipping the pre-swelled polymer film (chloroform, 24 h) in a chloroform solution of CX (5% and 10% concentration) for 24 h at 25 °C for both HDPE and PP.

Reflectance (%) spectra (Figure 7a,a') and TGA analysis (Figure 7b,b',c,c') were almost unchanged in HDPE samples, while they clearly evidenced the inclusion of CX in the PP polymer. A band at 230 nm and 276 nm and a peak at 348 °C relative to CX absorption (Figure 7a') and degradation (Figure 7b',c'), respectively, were observed. According to TGA, about 2% of the calixarene derivative included in PP was determined for the sample dipped in the 5% CX solution.

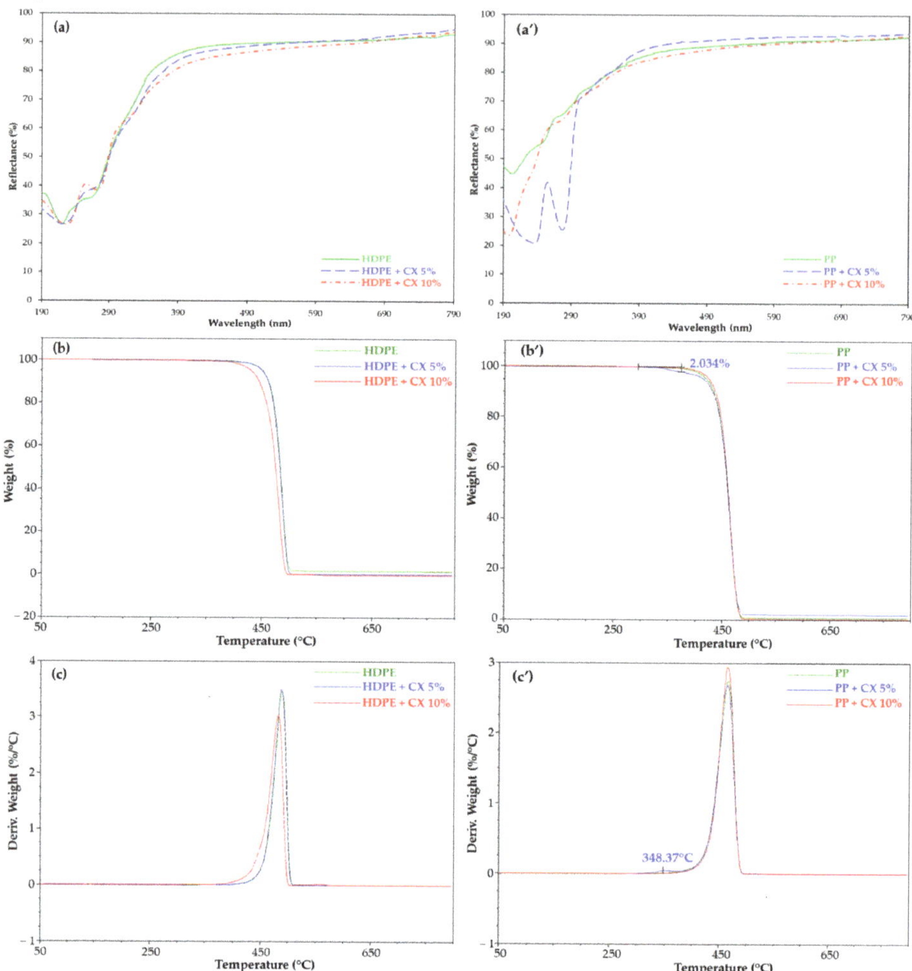

Figure 7. Reflectance (%) spectra, TGA, and DTG of (**a–c**) HDPE and (**a′–c′**) PP films, empty (green line) and embedded with CX (blue and red lines).

The higher affinity of the CX for PP than the HDPE film could be related to the presence of CH_3 groups in PP that can establish interactions with the CX aromatic rings as well as to the higher swelling index in $CHCl_3$ (Figure S10). The unchanged values of the SCA in HDPE embedded with CX at room temperature further confirmed the lower inclusion of CX in the HDPE matrix. Thus, successive tests were focused on the PP polymer matrix, using a 5% CX solution.

Two procedures were carried out to incorporate the CX/iodine complex in PP: a one-step inclusion of the complex and a two-step incorporation, including CX at first and then forming its complex through the exposure of the CX-loaded PP samples to iodine vapor or water solution. A comparative test with iodine-loaded PP samples was carried out through the exposure of the polymer matrices ("empty PP") to I_2 vapor or water solution.

For embedding iodine into the CX-loaded or empty PP, the polymer films were dipped into a water solution of iodine (1.18×10^{-3} M) or exposed to iodine vapor for 24 h at room temperature. The embedding of iodine in the polymer films was evident with the naked eye from the dark orange coloration and confirmed by reflectance spectra that showed the typical signals of iodine (Figure 8).

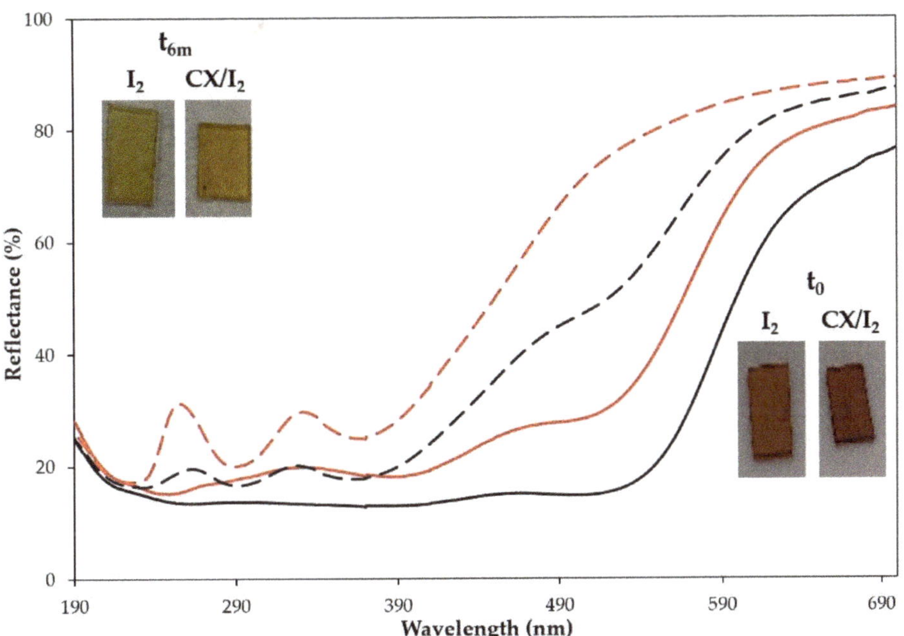

Figure 8. Reflectance (%) spectra of PP polymer films loaded with iodine (red lines) or CX/iodine complex by vapor method (black lines) at t_0 (full lines) and after 6 months (dashed lines). Insets: pictures of the samples at t_0 and after 6 months.

It was evident by monitoring the samples over time that the adsorbed iodine was lost from all samples, but a higher amount of I_2 was retained in the CX-loaded polymers (Figure 8 and insets). Similar results were obtained when the empty films and CX-loaded PP entrapped iodine by treatment with iodine vapors (Figure 8 and insets).

We also investigated the one-step embedding of the CX/iodine complex into PP films. To these ends, the polymer films, pre-swelled in chloroform, were dipped in a chloroform solution of the CX/iodine complex or iodine as control (7.5 mg/mL chloroform). After solvent evaporation, the polymer films with and without CX showed a different coloration and reflectance spectra (Figure 9, inset). Reflectance spectra confirmed the embedding of iodine and CX/iodine in the PP films. After 6 months, no significant difference in the reflectance spectra of CX/iodide-loaded PP film (Figure 9) was observed, evidencing a good stability of the sample, which is relevant to the evaluation of an eventual barrier effect over time.

Changes in the wettability of the CX-loaded films, consistent with the presence of iodine, were observed as well (Figure 10). In Figure 10, the contact angle values for the embedded PP (Figure 10) samples and unmodified are compared. For PP dipped in the CX solution at 5%, the SCA value is much higher than that of the unmodified polyolefin. This agrees with the lower wettability due to the CX hydrophobicity. The SCA of polyolefin film portions loaded with CX/iodine complex decreased in agreement with the presence of the iodine.

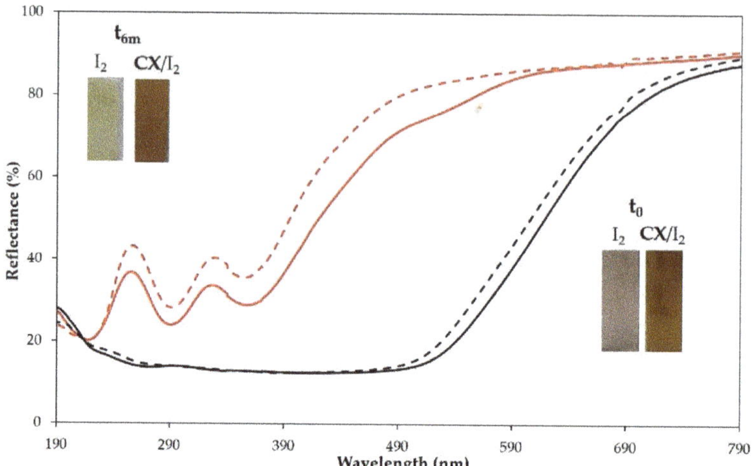

Figure 9. Reflectance (%) spectra of PP polymer films loaded with iodine (red lines) or CX/iodine complex in chloroform solution (black lines) at t_0 (full lines) and after 6 months (dashed lines). Insets: pictures of the samples at t_0 and after 6 months.

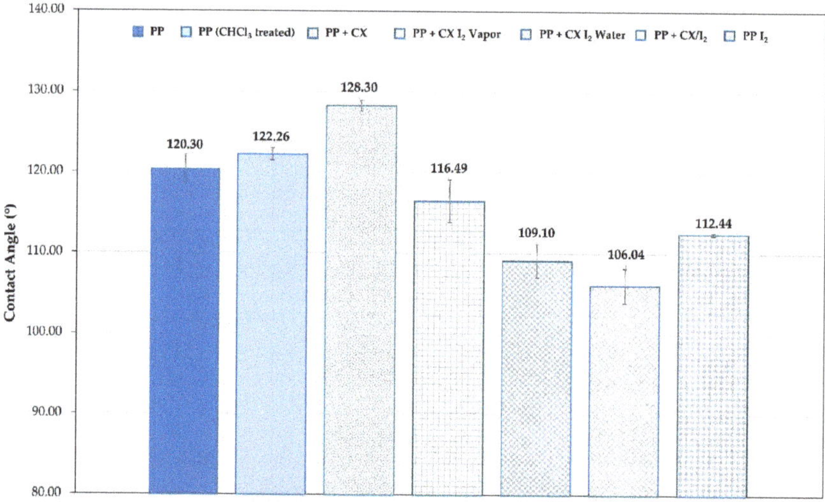

Figure 10. SCA of PP strip samples unmodified and embedded with iodine, CX, or its iodine complex.

Figure 11 shows the TGA and DTG of the PP strip samples unmodified and embedded with CX and its iodine complex. Noteworthy, TGA highlighted the presence of a degradative step related to the CX/iodine included and allowed the determination of the percentage in the film samples prepared by single- and two-step approaches (Figure 11a). A higher percentage was detected for the PP strip with the CX/iodine complex embedded via a single step.

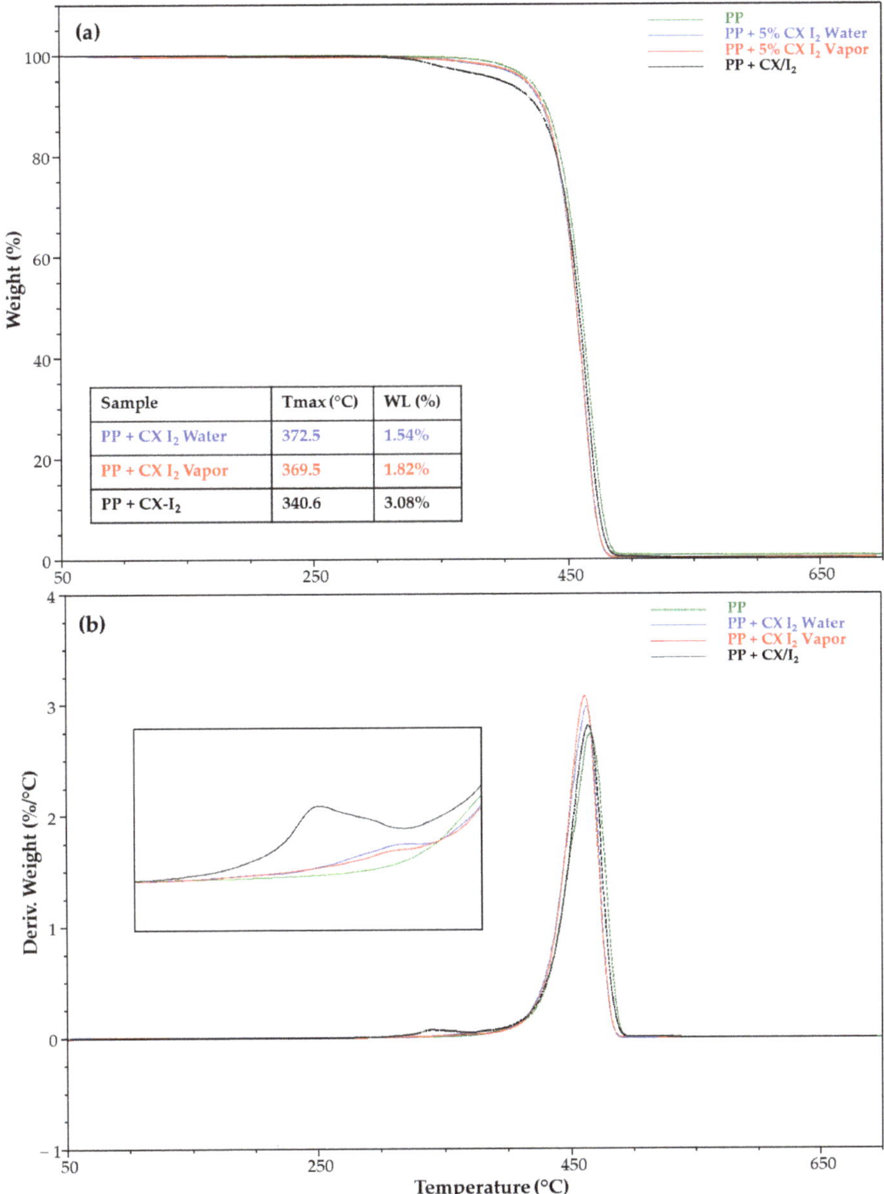

Figure 11. (**a**) TGA and (**b**) DTG of PP strip samples unmodified and embedded with CX iodine complex.

2.5. Influence of Calix[4]arene Derivative-Polypropylene Systems on Leaching of Iodine through Packaging from a PVP-I Buffered Solution

A comparative test was carried out to check the influence of CX-polyolefin systems on the leaching of iodine through packaging from a PVP-I solution (0.3% in buffer, pH = 6) simulating a low-concentration iodine formulation. The I_2 present in the PVP-I solution was determined by extraction with cyclohexane which, like other organic water-immiscible solvents, removes only the non-ionic iodine (I_2) and not the anionic iodine species present in PVP-I solutions. The amount of iodine extracted was determined by UV-vis spectroscopy. To simulate the plastic dropper of a container, we replaced the Teflon septum in a gas chromatography vial with PP circular film samples made in the laboratory. To evaluate the barrier effect, comparative tests were carried out with vials containing 0.8 mL of PVP-I phosphate citrate buffered solution (pH 6) and with a septum of circle-shaped PP samples, unmodified and embedded with CX and its iodine complex, as well as I_2. The samples underwent accelerated testing at 40 °C for up to 7 days. The amount of iodine present in the PVP-I solution, at time = 0 and after 7 days at 40 °C, was determined by cyclohexane extraction. In Figure 12, the residual iodine (%) in the PVP-I buffered solution is reported. The iodine amount in solution decreases for all the samples. However, the residual I_2 (%) in the PVP-I solution covered by Teflon and the circle-shaped unmodified PP is lower than that found for PP samples embedded with CX and its complexes (PP + CX, PP + CX I_2 vapor, and PP + CX/I_2, Figure 12). This behavior highlights higher iodine leaching in the absence of CX. Reasonably, PP + CX absorbs I_2 vapor from the PVP-I buffered solution and, by reproducing the preparation of the PP + CX I_2 vapor samples, provides a more valuable barrier effect in comparison with the PP unmodified sample. As expected, the PP samples (PP + I_2 vapor and PP + I_2 water, Figure 12) embedded with iodine, without CX, also show a higher residual I_2 (%) than unmodified PP, rather similar to Teflon.

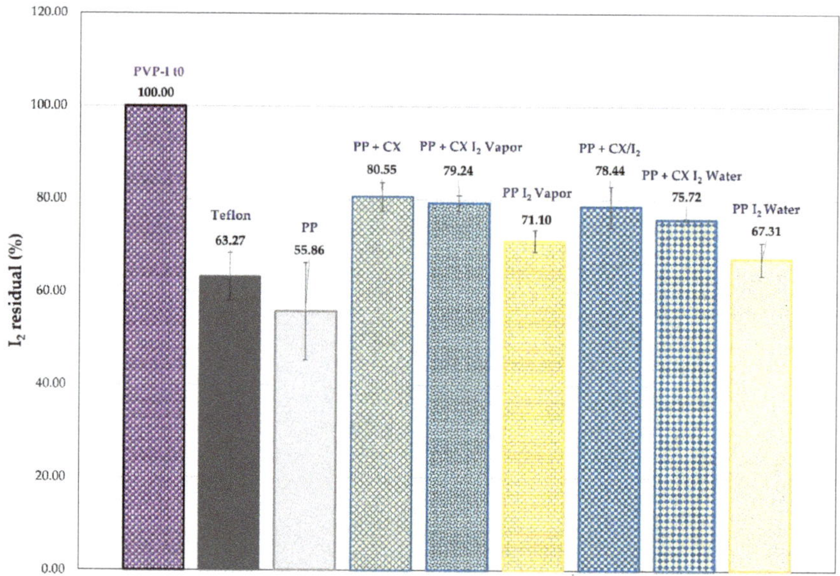

Figure 12. Residual iodine (%) determined in PVP-I buffered solution after 7 days at 40 °C in vials closed with cups of PP samples, unmodified and embedded with CX and its complex and I_2.

Reflectance spectra of PP circular film samples, unmodified and loaded with iodine or CX/iodine complex, at t = zero and after 7 days at 40 °C, are reported in Figure 13. The spectra of PP and PP + CX after 7 days clearly highlight a decrease in reflectance (%) in the iodine absorption region, indicative of the absorption of iodine vapor from the PVP-I buffered solution, which is also supported by the yellow coloring of the samples. On the contrary, PP embedded with iodine shows an increase in reflectance and coloring after 7 days at 40 °C due to iodine leaching. Remarkably, no significant variation in reflectance is observed in PP embedded with the CX/iodine complex, since CX successfully retains iodine in its cavity. In agreement with the results revealed in the extraction of residual iodine (Figure 12) from the PVP-I buffered solution, the iodine complexed in the CX provides the more effective barrier effect.

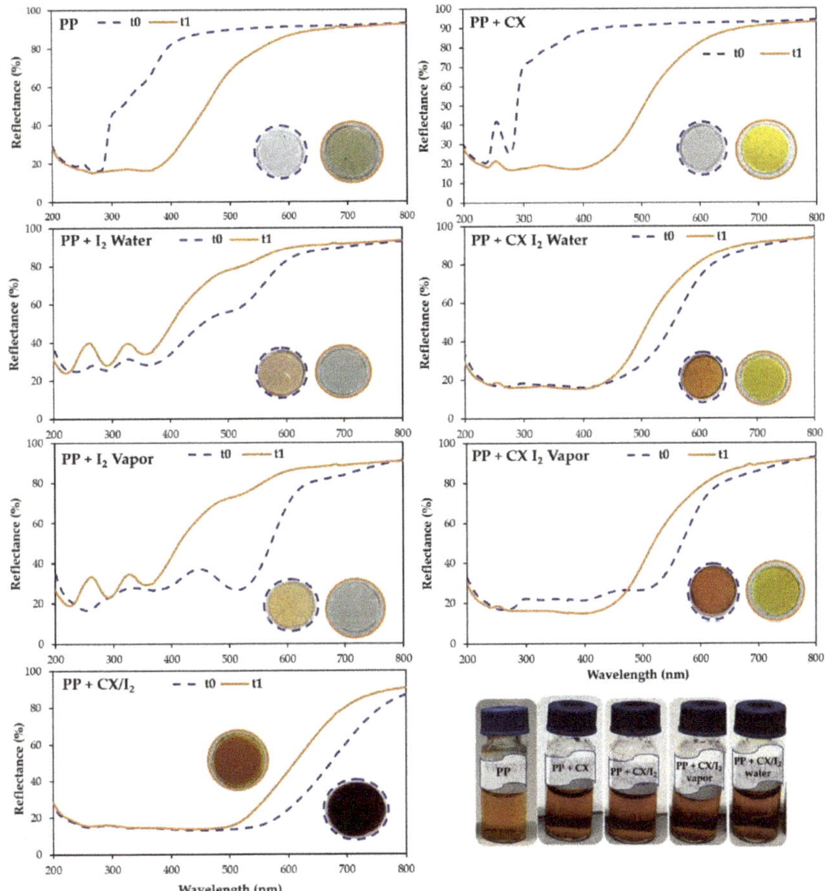

Figure 13. Reflectance (%) spectra and photographic documentation of PP circular film samples unmodified and loaded with iodine or CX/iodine complex at t = 0 (dashed blue line) and after 7 days placed on the cup of a 0.3% PVP-I water solution (solid orange line) at 40 °C.

3. Materials and Methods

3.1. Materials

All reagents and solid iodine were purchased from Sigma-Aldrich Chemical Co., Ltd. (Milan, Italy). PVP-I was obtained from BASF (Milano, Italy). HDPE (PHARMALENE® MP 90 PH, Versalis, Milano, Italy) and PP (Purell HP371P, Lyondellbasell, Milano, Italy) were used for film preparation. Tetra(ethoxycarbonyl-methoxy)-4-tert-butylcalix[4]arene (CX) was prepared as reported in the literature [23].

3.2. Preparation of CX/Iodine Complex in Chloroform

CX was added to a chloroform iodine solution (1×10^{-3} M) in different molar ratios (from 1:0.5 to 1:3). The mixture was stirred for 15 min at room temperature and analyzed by UV-vis spectroscopy.

3.3. Preparation of CX/Iodine Complex from Iodine Water Solution

The water-insoluble powder of CX (200 mg) was added to 200 mL of a 0.3 mg/mL iodine water solution and stirred at room temperature for 24 h. The orange powder of the complex was recovered by filtration, washed with pure water, and dried.

3.4. Preparation of CX/Iodine Complex from Iodine Vapors

A magnetically stirred powder of CX was placed in a closed container under iodine vapors at 75 °C. The complex formation was monitored over time.

3.5. Preparation of Polyolefin Films

Polyolefin films (average thickness 500 μm) were obtained by a hot press machine PM 20/200 (DGTS srl, Verduggio, Monza Brianza, Italy). Polymer pellets (1.5–2 g) were compressed for 2 min under a pressure of 25–30 bar between two Teflon plates (thickness: 500 μm) containing a spacer (thickness: 500 μm) at 190 °C for HDPE and 210 °C for PP and subsequently water-cooled. Rectangular (0.7 × 4 cm) and circular (0.8 cm) film samples for the different tests were obtained from a cutter. The hot-pressed films were stored at room temperature for at least three weeks before use to reach equilibrium crystallinity.

3.6. Embedding of CX/Iodine Complex in HDPE/PP Samples

HDPE/PP samples were pre-swelled in chloroform at room temperature overnight to remove impurities, and then they were dipped in a 5% chloroform solution of the CX/iodine complex prepared by the iodine water solution method and placed in a shaker for 24 h.

3.7. Embedding of CX in HDPE/PP Samples

HDPE/PP samples, pre-swelled in chloroform, were dipped in a 5% chloroform solution of CX (5% or 10%). The sample was placed in a shaker overnight and then dried in an oven at 40 °C.

3.8. Loading of Iodine in the CX-Embedded PP Samples

The loading of iodine into PP samples embedded with CX was performed by two methods: (A) dipping in a water solution of iodine (0.33 mg/mL) for 24 h and (B) contact with iodine vapor for 24 h at 40 °C.

3.9. Nuclear Magnetic Resonance (NMR)

The ^{1}H NMR spectrum of CX was acquired on a Bruker Avance 400 spectrometer (400 MHz, $CDCl_3$, 297 K). The spectral signals (Figure S1) were consistent with data reported in the literature [23].

3.10. UV-Vis Spectroscopy

UV-vis spectra were recorded at room temperature on an Agilent Technologies 8453 UV-Vis spectrophotometer using quartz cuvettes with a path length of 1 cm and Jasco v770 spectrophotometer equipped with an integrating sphere for reflectance measurements.

3.11. Thermogravimetric Analysis (TGA)

Thermogravimetric analyses (TGA) were performed using a thermogravimetric apparatus TGA Q500 (TA instruments, New Castle, DE, USA) under a nitrogen atmosphere at a 10 °C/min heating rate, from 50 °C to 800 °C. Sample weight was approximately 5 mg. The weight loss percent and its derivative (DTG) were recorded as a function of temperature.

3.12. Contact Angle Measurements

The surface wettability values of samples were measured at room temperature using a contact angle goniometer (OCA15EC, Dataphysics, Filderstadt, Germany). Static contact angle (SCA) values were determined by dropping 2 μL of water from a micro syringe onto the surfaces and analyzing the images taken by the connected video camera with software (SCA 20). To eliminate interference, the samples were previously equilibrated for 30 min at 40 °C and then SCA was measured. At least five measurements were carried out for each sample to ensure repeatability of the experiments.

3.13. Determination of Leaching of Iodine through CX-Loaded PP Films

The septum in gas chromatography vials was replaced with PP circular film samples made in the laboratory. A comparative test was carried out to check the influence of CX-polyolefin systems on the leaching of iodine through the packaging from a PVP-I solution (0.3% in phosphate citrate buffer, pH = 6). The I_2 present in the PVP-I solution was determined by extraction with cyclohexane. The amount of iodine extracted was determined by UV-vis spectroscopy. To evaluate the barrier effect, comparative tests were carried out with vials containing 0.8 mL of PVP-I buffered solution and modified with PP circle-shaped samples, unmodified and embedded with CX (PP + CX) and its iodine complex (PP + CX I_2 vapor, PP + CX I_2 water), and I_2 (PP + I_2 vapor, PP + I_2 water). Unmodified PP, PP embedded with CX, or CX-iodine complexes, otherwise known as iodine, and the original septum in Teflon, used for comparison, were used in triplicate. The samples underwent accelerated testing at 40 °C for up to 7 days. The amount of iodine present in the PVP-I solution at time = 0 and after 7 days at 40 °C was determined by UV-vis spectroscopy after extraction with cyclohexane (0.8 mL) at λ = 523 nm by referring to a calibration curve. Weight loss or increase in the PP circular film samples after 7 days was monitored. Reflectance (%) spectra were recorded at time = 0 and after 7 days at 40 °C.

4. Conclusions

The capability of a hydrophobic calix[4]arene derivative blocked in a cone conformation to complex iodine from vapor as well as water or chloroform solution was demonstrated by UV-vis spectroscopy. Reducing iodine loss from plastic containers is still a challenge. Thus, to contribute to the research of new approaches to reduce the loss of I_2 from an iodophor solution, the calix[4]arene/iodine complex was embedded in polyolefin (HDPE and PP) via a swelling procedure. PP showed a higher affinity than HDPE for the CX. The PP films (unmodified and iodine- and calixarene/iodine-loaded) were characterized for surface and bulk changes by contact angle, TGA, and UV-vis diffuse reflectance spectra. A comparative study showed that the films with CX/iodine complex more effectively reduce iodine leaching from a low-concentration PVP-I solution and successfully provide a higher barrier effect.

Overall, preliminary results highlight that calixarene-embedded polyolefins could be promising novel materials to enhance the shelf-life of iodine-based formulations. Further studies are underway for optimizing the embedding of CX and other iodophor calixarene derivatives in polyolefin films. This study paves the way for the development of novel calixarene/polyolefin blends with potential applicability in different fields, including the development of novel materials with antiseptic properties.

Supplementary Materials: The following supporting information can be downloaded at: https: //www.mdpi.com/article/10.3390/molecules28041869/s1, Figure S1: ^{1}H NMR spectrum of the tetra(ethoxycarbonyl-methoxy)-4-*tert*-butylcalix[4]arene (CX) (400 MHz, CDCl$_3$, 297 K); Figure S2: UV-vis spectra of iodine CHCl$_3$ solution (1.3×10^{-3} M) in the presence of increasing amounts of CX (from 0 to 3.9×10^{-3} M); Figure S3: Job's plot for the determination of the stoichiometry of the CX/iodine complex; Figure S4: Plot of absorbance change versus mole ratio [CX]/[Iodine] at λ = 366 nm; Figure S5: Benesi-Hildebrand plot of absorption data for the CX/iodine complex; Figure S6: UV-vis spectra of iodine in water solution (1.3×10^{-3} M) alone and after stirring with increasing amounts of CX powder (iodine:CX 1:2, 1:6, and 1:10 molar ratio); Figure S7: UV-vis spectra of (**a**) iodine in water solution (1.3×10^{-3} M) before (blue line) and after stirring with CX powder (iodine:CX, 2:1 molar ratio) (orange line); (**b**) iodine extracted by cyclohexane from the water solution of iodine (1.3×10^{-3} M) alone (blue line) and after stirring with CX powder (iodine:CX, 2:1 molar ratio) (orange line); Figure S8: UV-vis spectrum in chloroform of the CX/iodine complex from solid-air method; Figure S9: (**a**) TGA and (**b**) isothermal at 50 °C (60 min) of I_2 sample; Figure S10: Swelling index in CHCl$_3$ vs. time for HDPE and PP strip samples at room temperature.

Author Contributions: Conceptualization, P.R. and G.M.L.C.; methodology, L.F., M.R., M.L., C.M., G.M.L.C., P.R. and E.T.A.S.; validation, E.T.A.S., P.R. and G.M.L.C.; investigation, L.F., M.R., M.L., C.M., G.M.L.C., P.R. and E.T.A.S.; resources, E.T.A.S., P.R., N.D. and G.M.L.C.; data curation, L.F., M.R., M.L., C.M. and E.T.A.S.; writing—original draft preparation, E.T.A.S., P.R. and G.M.L.C.; writing—review and editing, L.F., M.R., M.L., C.M., N.D., G.M.L.C., P.R. and E.T.A.S.; visualization, L.F., M.R., M.L., C.M., G.M.L.C., P.R. and E.T.A.S.; supervision, E.T.A.S., P.R. and G.M.L.C.; project administration, P.R. and G.M.L.C.; funding acquisition, N.D., P.R. and G.M.L.C. All authors have read and agreed to the published version of the manuscript.

Funding: This research was funded by PO FESR 2014–2020, Action 1.1.5, project "New therapeutic strategies in ophthalmology: bacterial, viral and microbial infections-NUSTEO", CUP: G68I18000700007-application code 08CT2120090065.

Institutional Review Board Statement: Not applicable.

Informed Consent Statement: Not applicable.

Data Availability Statement: The data presented in this study are available on request from the corresponding authors.

Acknowledgments: The authors thank BASF for the "PVP-I" sample. Many thanks are due to Roberto Rapisardi (CNR IPCB Catania, Italy) and Agatino Renda (CNR ICB Catania, Italy), for their skillful technical assistance and to Maria Serenella Vitale (CNR IPCB Catania, Italy) and Provvidenza Guagliardo (CNR IPCB Catania, Italy) for continuous administrative support.

Conflicts of Interest: The authors declare no conflict of interest. The funders had no role in the design of the study; in the collection, analyses, or interpretation of data; in the writing of the manuscript; or in the decision to publish the results.

References

1. Gustsche, C.D. *Calixarenes Revisited*; The Royal Society of Chemistry: Cambridge, UK, 1998.
2. Mandolini, L.; Ungaro, R. *Calixarenes in Action*; Imperial College Press: London, UK, 2000.
3. Bohmer, V. Calixarenes, Macrocycles with (Almost) Unlimited Possibilities. *Angew. Chem. Int. Ed. Engl.* **1995**, *34*, 713–725. [CrossRef]
4. Taghvaei-Ganjali, S.; Khosravi, M.; Tahvildari, K.; Naderi, A. Spectroscopic studies of interaction between Iodine and Benzyloxy ether Derivatives of Calix[4]arene. *J. Sci. Islam. Azad Univ.* **2006**, *16*, 6–11.
5. Mizyed, S.A.; Ashram, M.; Saymeh, R.; Marji, D. A Thermodynamic study of the charge transfer complexes of iodine with different tert-butylcalix[4]crowns. *Z. Naturforsch.* **2005**, *60b*, 1133–1137. [CrossRef]
6. An, D.; Li, L.; Zhang, Z.; Asiri, A.M.; Alamry, K.A.; Zhang, X. Amino-bridged covalent organic polycalix[4]arenes for ultra efficient adsorption of iodine in water. *Mat. Chem. Phys.* **2020**, *239*, 122328. [CrossRef]
7. Shetty, D.; Raya, J.; Han, D.S.; Asfari, Z.; Olsen, J.-C.; Trabolsi, A. Lithiated polycalix[4]arenes for efficient adsorption of iodine from solution and vapor phases. *Chem. Mater.* **2017**, *29*, 8968–8972. [CrossRef]
8. Sharma, P.R.; Pandey, S.; Soni, V.K.; Choudhary, G.; Sharma, R.K. Macroscopic recognition of iodide by polymer appended calix[4]amidocrown resin. *Supramol. Chem.* **2019**, *31*, 634–644. [CrossRef]
9. Zhang, Z.; Li, L.; An, D.; Li, H.; Zhang, X. Triazine-based covalent organic polycalix[4]arenes for highly efficient and reversible iodine capture in water. *J. Mater. Sci.* **2020**, *55*, 1854–1864. [CrossRef]
10. Skorjanc, T.; Shetty, D.; Trabolsi, A. Pollutant removal with organic macrocycle-based covalent organic polymers and frameworks. *Chem* **2021**, *7*, 882–918. [CrossRef]
11. Su, K.; Wang, W.; Li, B.; Yuan, D. Azo-Bridged Calix[4]resorcinarene-Based Porous Organic Frameworks with Highly Efficient Enrichment of Volatile Iodine. *ACS Sustain. Chem. Eng.* **2018**, *6*, 17402–17409. [CrossRef]
12. Moulay, S. Molecular iodine/complexes. *J. Polym. Eng.* **2013**, *33*, 389–443. [CrossRef]
13. Koerner, J.C.; George, M.J.; Meyer, D.R.; Rosco, M.G.; Habib, M.M. Povidone-iodine concentration and dosing in cataract surgery. *Surv. Ophthalmol.* **2018**, *63*, 862–868. [CrossRef] [PubMed]
14. Siggia, S. The chemistry of polyvinylpyrrolidone-iodine. *J. Am. Pharm. Assoc.* **1957**, *46*, 201–204. [CrossRef]
15. Bhagwat, D.; Oshlack, B. Stabilized PVP-I Solutions. U.S. Patent 5,126,127, 30 June 1992.
16. Isenberg, S.J.; Apt, L.; Yoshimori, R.; Pham, C.; Lam, N.K. Efficacy of topical Povidone-Iodine during the first week after ophthalmic surgery. *Am. J. Ophthalmol.* **1997**, *124*, 31–35. [CrossRef] [PubMed]
17. Bhagwat, D.; Iny, O.; Pedi, F., Jr. Stabilizing Packaged Iodophor and Minimizing Leaching of Iodine through Packaging. EP Patent 0371283 A2, 6 June 1990.
18. Zhu, W.; Gou, P.; Shen, Z. Applications of calixarenes in polymer synthesis. *Macromol. Symp.* **2008**, *261*, 74–84. [CrossRef]

19. Zheng, Y.-S.; Ying, L.-Q.; Shen, Z.-Q. Polymerization of propylene oxide by a new neodymium complex of calixarene derivative. *Polymer* **2000**, *41*, 1641–11643. [CrossRef]
20. Feng, W.; Yuan, L.H.; Zheng, S.Y.; Huang, G.L.; Qiao, J.L.; Zhou, Y. The effect of p-tert-butylcalix[n]arene on γ-radiation degradation of polypropylene. *Radiat. Phys. Chem.* **2000**, *57*, 425–429. [CrossRef]
21. Mattiuzzi, A.; Troian-Gautier, L.; Mertens, J.; Reniers, F.; Bergamini, J.-F.; Lenne, Q.; Lagrost, C.; Jabin, I. Robust hydrophobic gold, glass and polypropylene surfaces obtained through a nanometric covalently bound organic layer. *RSC Adv.* **2020**, *10*, 13553–13561. [CrossRef] [PubMed]
22. Chennakesavulu, K.; Raviathul Basariya, M.; Sreedevi, P.; Raju, G.B.; Prabhakar, S.; Rao, S.S. Study on thermal decomposition of calix[4]arene and its application in thermal stability of polypropylene. *Thermochim. Acta* **2011**, *515*, 24–31. [CrossRef]
23. Arnaud-Neu, F.; Collins, E.M.; Deasy, M.; Ferguson, G.; Harris, S.J.; Kaitner, B.; Lough, A.J.; McKervey, M.A.; Marques, E.; Ruh, B.L.; et al. Synthesis, X-ray crystal structures, and cation-binding properties of alkyl calixaryl esters and ketones, a new family of macrocyclic molecular receptors. *J. Am. Chem. Soc.* **1989**, *111*, 8681–8691. [CrossRef]

Article

Critical Analysis of Association Constants between Calixarenes and Nitroaromatic Compounds Obtained by Fluorescence. Implications for Explosives Sensing

Alexandre S. Miranda [1,2], Paula M. Marcos [1,3,*], José R. Ascenso [4], Mário N. Berberan-Santos [2,*], Peter J. Cragg [5], Rachel Schurhammer [6] and Christophe Gourlaouen [7]

[1] Centro de Química Estrutural, Institute of Molecular Sciences, Faculdade de Ciências, Universidade de Lisboa, Edifício C8, 1749-016 Lisboa, Portugal

[2] IBB-Institute for Bioengineering and Biosciences, Instituto Superior Técnico, Universidade de Lisboa, 1049-001 Lisboa, Portugal

[3] Faculdade de Farmácia da Universidade de Lisboa, Av. Prof. Gama Pinto, 1649-003 Lisboa, Portugal

[4] Centro de Química Estrutural, Institute of Molecular Sciences, Instituto Superior Técnico, Complexo I, Av. Rovisco Pais, 1049-001 Lisboa, Portugal

[5] School of Applied Sciences, Huxley Building, University of Brighton, Brighton BN2 4GJ, UK

[6] Laboratoire de Modélisation et Simulations Moléculaires, Université de Strasbourg, UMR 7140, F-67000 Strasbourg, France

[7] Laboratoire de Chimie Quantique, Université de Strasbourg, UMR 7177, F-67000 Strasbourg, France

[*] Correspondence: pmmarcos@fc.ul.pt (P.M.M.); berberan@tecnico.ulisboa.pt (M.N.B.-S.)

Abstract: The binding behaviour of two ureido-hexahomotrioxacalix[3]arene derivatives bearing naphthyl (**1**) and pyrenyl (**2**) fluorogenic units at the lower rim towards selected nitroaromatic compounds (NACs) was evaluated. Their affinity, or lack of it, was determined by UV-Vis absorption, fluorescence and NMR spectroscopy. Different computational methods were also used to further investigate any possible complexation between the calixarenes and the NACs. All the results show no significant interaction between calixarenes **1** and **2** and the NACs in either dichloromethane or acetonitrile solutions. Moreover, the fluorescence quenching observed is only apparent and merely results from the absorption of the NACs at the excitation wavelength (inner filter effect). This evidence is in stark contrast with reports in the literature for similar calixarenes. A naphthyl urea dihomooxacalix[4]arene (**3**) is also subject to the inner filter effect and is shown to form a stable complex with trinitrophenol; however, the equilibrium association constant is greatly overestimated if no correction is applied (9400 M^{-1} vs 3000 M^{-1}), again stressing the importance of taking into account the inner filter effect in these systems.

Keywords: hexahomotrioxacalix[3]arenes; nitroaromatic compounds; calixarene-nitroaromatic supramolecular association; explosives sensing; fluorescence inner filter effect

Citation: Miranda, A.S.; Marcos, P.M.; Ascenso, J.R.; Berberan-Santos, M.N.; Cragg, P.J.; Schurhammer, R.; Gourlaouen, C. Critical Analysis of Association Constants between Calixarenes and Nitroaromatic Compounds Obtained by Fluorescence. Implications for Explosives Sensing. *Molecules* **2023**, *28*, 3052. https://doi.org/10.3390/molecules28073052

Academic Editor: Takahiro Kusukawa

Received: 3 March 2023
Revised: 24 March 2023
Accepted: 27 March 2023
Published: 29 March 2023

1. Introduction

The detection of explosives continues to be an imperative task for all countries in their antiterrorist and homeland security activities. These materials, with their huge destruction potential, can be cheaply and easily prepared. Among them, nitroaromatic compounds (NACs), such as trinitrotoluene (TNT), dinitrotoluene (DNT) and trinitrophenol (TNP, also known as picric acid), are common explosives used for military purposes and the principal components of landmines [1,2]. They are also employed in agrochemical and pharmaceutical industries, being considered environmental pollutants. Concerning human health, they can cause skin irritation, eye cataracts, male infertility, kidney and liver damage, among other diseases [1,2].

Although a wide variety of methods have been used for trace detection of explosives, such as mass spectrometry, ion mobility spectrometry, surface enhanced Raman spectroscopy, gas chromatography coupled with different detectors and nuclear quadruple resonance, expensive and complex equipment is required, causing limitations for their use in real-time applications [3]. Therefore, low-cost detection techniques, with high portability, high sensitivity and selectivity are needed for in-field analyte effective sensing. Among analytical techniques, luminescence-based methods meet these requirements [4].

Lately, a wide range of fluorescence sensors for monitoring explosives in the solid, solution and vapor phases have been developed [5], in particular based on calixarenes [6,7]. Owing to their structural features, these macrocycles have been widely investigated as ion and neutral molecule receptors [8,9]. They possess a pre-organized cavity available in different sizes and conformations, and they can be functionalised at the upper and lower rims, leading to an almost unlimited number of derivatives. Fluorophores such as naphthalene, anthracene and pyrene are among the most incorporated in the calixarene framework, leading to potential fluorescent probes for NACs [10–16]. The introduction of calix[4]arene moieties in fluorescent conjugated polymers has also been used to detect explosives [17,18].

In the course of our recent studies on anion binding by fluorescent homooxacalixarenes [19–21], calixarene analogues in which the CH_2 bridges are partly or completely replaced by CH_2OCH_2 groups [22,23], we have extended our research into the recognition of a different kind of analyte, namely nitroaromatic compounds, by hexahomotrioxacalix[3]arene-based receptors. These macrocyclic compounds, with an 18-membered ring and having only two basic conformations, have already been investigated to detect explosives. A triazole-modified hexahomotrioxacalix[3]arene has been reported as a selective chemosensor for TNP, evidenced by UV-Vis and fluorescence studies [24]. Thus, in this paper, we investigate the binding behaviour of two ureido-hexahomotrioxacalix[3]arene derivatives recently obtained and containing naphthyl (**1**) or pyrenyl (**2**) residues at the lower rim, towards selected NACs (Scheme 1). Their affinity, or lack of it, was determined by UV-Vis absorption, fluorescence and NMR spectroscopy. Different computational methods were also performed to bring further insights to the analysis of the complex formation between the calixarenes and the NACs.

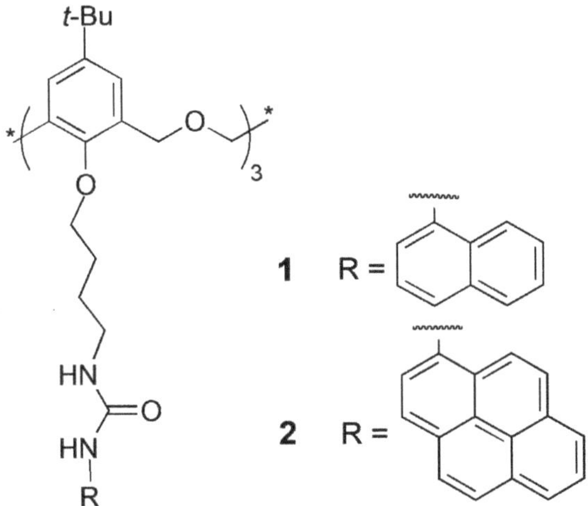

Scheme 1. Chemical structures of compounds **1** and **2**.

2. Results and Discussion

2.1. Detection of Nitroaromatic Explosives

2.1.1. UV-Vis Absorption and Fluorescence Studies

Recently, we reported the anion binding properties of fluorescent ureido-hexahomotrioxacalix[3]arene receptors **1** and **2**, obtained in the partial cone conformation [21]. Following this line of research, the potential of naphthyl (**1**) and pyrenyl (**2**) ureas as chemosensors to detect nitroaromatic explosives was explored by fluorescence titrations. Selected NACs, 2-nitrotoluene (NT), 2,4-dinitrotoluene (DNT), 2,4,6-trinitrotoluene (TNT), 2,4,6-trinitrophenol (TNP), 1-nitrobenzene (NB) and 1,3-dinitrobenzene (DNB), were tested in dichloromethane and, in the case of **1**, also in acetonitrile (Figure 1). TNT from a military grenade was also analysed, showing similar results to the commercially sourced sample.

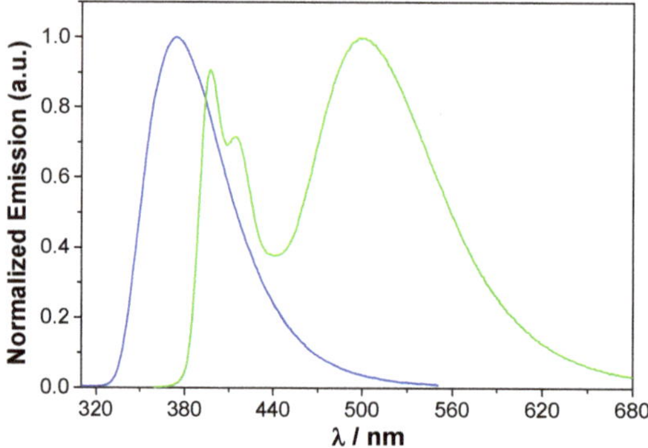

Figure 1. Structures of the nitroaromatic compounds studied.

Compound **1** exhibits emission bands at approximately 375 nm in dichloromethane and acetonitrile, characteristic of the naphthylurea group, while compound **2** presents both monomer (ca. 400 nm) and excimer (ca. 500 nm) bands in dichloromethane arising from the pyrene moieties, as reported previously (Figure 2) [21].

Figure 2. Normalised emission spectra of **1** (blue) and **2** (green) in CH_2Cl_2 at 25 °C. [**1**] = 20 μM, λ_{ex} = 300 nm; [**2**] = 10 μM, λ_{ex} = 340 nm.

A decrease in the fluorescence intensity was observed upon addition of all the NACs to **1** in both solvents, as shown for NB in Figure 3, as well as upon addition of TNP to **2** in dichloromethane (Figure S1).

As all NAC studied significantly absorb at 300 nm and TNP absorbs also at 340 nm (Figure 4 and Figure S2) the excitation wavelengths used in the case of naphthyl urea **1** and pyrenyl urea **2**, respectively, it is essential to correct for this inner filter effect (absorption of the NAC at the excitation wavelength) before concluding on the existence of genuine static fluorescence quenching (association quenching).

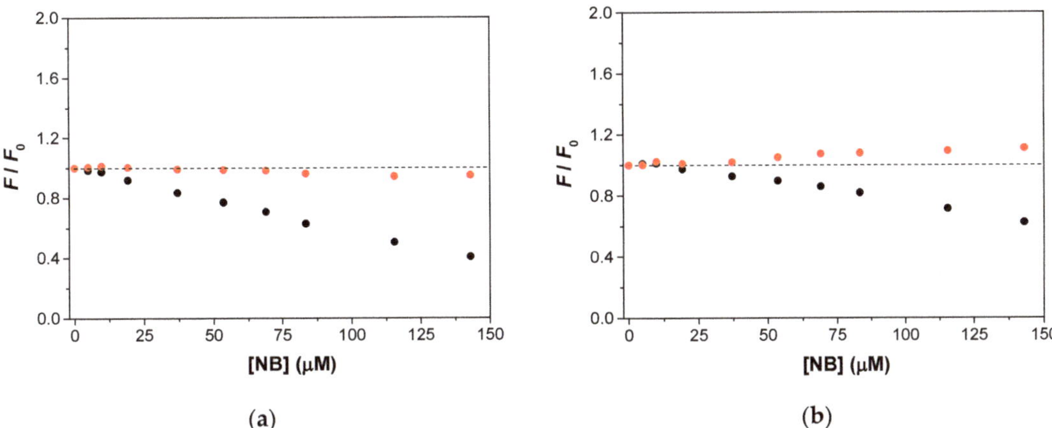

Figure 3. Relative fluorescence intensity (black circles) of Napht urea **1** (20 μM) upon addition of NB (up to 20 equiv) in (**a**) CH_2Cl_2 and (**b**) MeCN. F_0 is the intensity in the absence of quencher. Also shown are the intensities corrected for the inner filter effect (red circles). $\lambda_{ex} = 300$ nm.

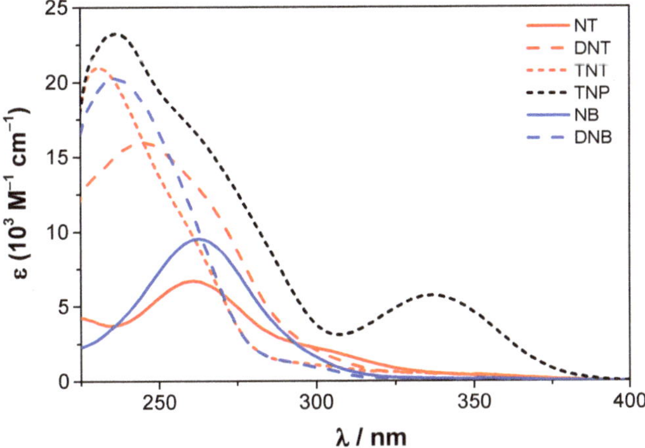

Figure 4. Electronic absorption spectra of the NACs in CH_2Cl_2.

In the case of emission collected only from the centre of the cell, as is the current situation, the correction of the excitation inner filter effect is straightforward, as it immediately follows on from Beer's law (1):

$$F_{corr} = 10^{\Delta A/2} F \tag{1}$$

where F_{corr} is the corrected fluorescence intensity, F the measured fluorescence intensity, and $\Delta A = A - A_0$, where A and A_0 are the absorbances (for 1 cm pathlength) at the excitation wavelength for the solutions of calixarenes **1** and **2** with and without NACs, respectively. The factor of $1/2$ in the exponent accounts for the 0.5 cm effective pathlength of the exciting beam.

As shown in Figures 3 and 5 (see also Figures S3–S7 for the remaining nitroaromatics), it is concluded that the fluorescence quenching by the NACs is only apparent and merely results from the excitation inner filter effect. The apparent quenching indeed correlates with the absorption coefficient of the NAC (Figure 4). No significant association thus exists between calixarenes **1** and **2**, and the NACs studied, both in dichloromethane and acetonitrile.

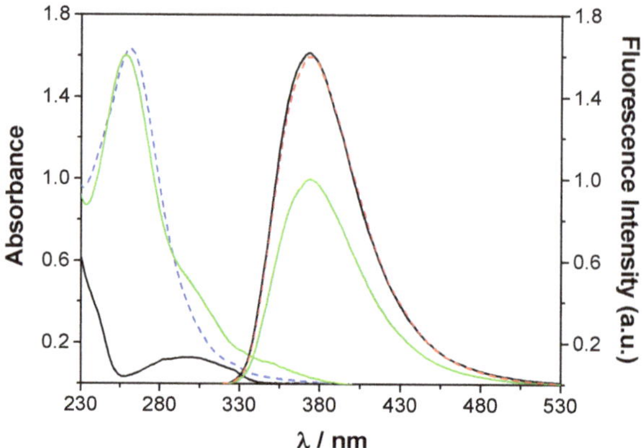

Figure 5. Absorption and fluorescence in CH_2Cl_2. Left side (short wavelengths): absorption spectra of **1** (10 µM) alone (black) and in the presence of 20 equiv. of NB (green). The absorption spectrum of 20 equiv. of NB is also shown (dashed blue). Right side (long wavelengths): fluorescence spectra ($\lambda_{ex} = 300$ nm) of **1** (10 µM) alone (black) and in the presence of 20 equiv. of NB, before (green) and after (dashed red) correction for the inner filter effect.

To further support this conclusion, UV-Vis absorption titrations were carried out with both urea compounds and all the NACs (see, for example, Figure 6 for **1** + DNB and Figures S8–S14), in order to evaluate association constants from absorption data.

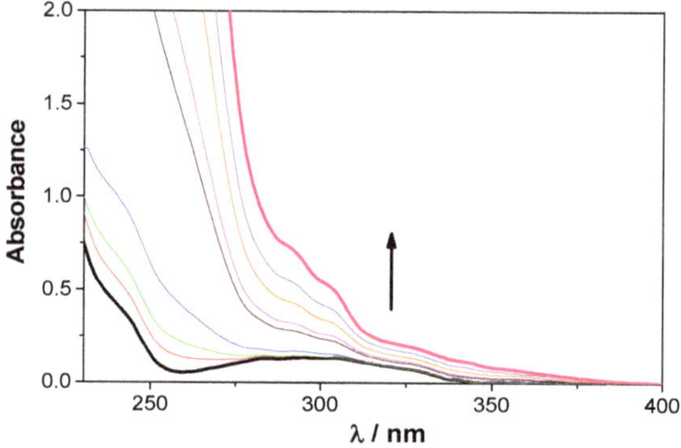

Figure 6. UV absorption spectra of Napht urea **1** (10 µM) upon addition of DNB (up to 30 equiv.) in CH_2Cl_2. The arrow indicates increasing amounts of DNB. The spectra of the mixtures are the result of the sum of the two constituent spectra.

The association constants determined (Table 1) for both **1** and **2** are indeed very small, always lower than 0.1 M^{-1}, with the exception of **1** + TNP in dichloromethane where it is slightly higher. Thus, the results obtained in all cases indicate only the existence of statistical complexes [25] of **1** and **2** with all NACs studied.

Table 1. Association constants (K_{ass}/M^{-1}) of ureas **1** and **2** with NACs determined by UV-Vis absorption at 25 °C.

	Solvent	NT	DNT	TNT	TNP	NB	DNB
1	CH_2Cl_2	<0.1	<0.1	<0.1	0.5	<0.1	<0.1
	MeCN	<0.1	<0.1	<0.1	<0.1	<0.1	<0.1
2	CH_2Cl_2	<0.1	<0.1	<0.1	<0.1	<0.1	<0.1

The fluorescence decays of **1** can be fitted with a sum of three exponentials only (Table 2). The complexity of the decay may result from the existence of several different configurations of this supramolecular entity. All three decay components (and hence the average lifetimes) show no significant changes (both in lifetimes and amplitudes) upon NACs addition in substantial amounts (up to 30 equiv.), as shown in Table 2. This invariability, associated with the apparent quenching of the steady-state fluorescence, could mean that static quenching was operative. However, given the negligible association constants found by UV-Vis absorption titrations, the conclusion is simply that no quenching, either static or dynamic, takes place. Indeed, for a possible dynamic quenching mechanism with diffusion control, NAC concentrations (always less than 200 μM) are too low for it to be observable.

Table 2. Two- and three-exponential analysis of fluorescence decays of **1** with 30 equiv. of NACs in CH_2Cl_2 at 25 °C.

	τ_1/ns (%)	τ_2/ns (%)	τ_3/ns (%)	$\bar{\tau}$/ns
1 *	0.35 (6)	2.6 (59)	9.3 (35)	4.8
1 + NT	0.33 (6)	2.5 (58)	8.6 (36)	4.6
1 + DNT	0.37 (6)	2.6 (59)	9.1 (35)	4.7
1 + TNT	0.39 (6)	2.7 (60)	9.4 (34)	4.8
1 + TNP	0.33 (6)	2.5 (57)	8.7 (37)	4.6
1 + NB	0.35 (6)	2.6 (58)	9.0 (36)	4.7
1 + DNB	0.35 (6)	2.6 (58)	8.8 (36)	4.7

* λ_{ex} = 300 nm.

The conclusions based on fluorescence are in stark contrast with the claims from two previous reports on NAC sensing using the fluorescence of calixarenes bearing naphthyl [13] and phenylethynylenyl moieties [14]. In these two cases, the excitation inner filter effect is very likely at play but is not mentioned. Both inner filters, excitation and emission, are frequently overlooked or entirely ignored in complexation studies based on fluorescence. Even when not as drastic as in the above-mentioned systems, these effects, if not accounted for, may result in grossly overestimated association constants [18].

An example of a homooxacalixarene system where this situation is observed is in the interaction of **3** (Scheme 2) with TNP, with data now presented and analysed. This system also exhibits a strong inner filter effect (Figure S15). As shown in Figure 7, the excitation inner filter correction is not enough to restore the intensities to their original value. It is thus concluded that a genuine quenching exists. Given the concentrations of quencher, dynamic collisional quenching can be ruled out. This is indeed borne out by lifetime measurements as displayed in Table 3. It can thus be concluded that the quenching is static, corresponding to the formation of an association complex.

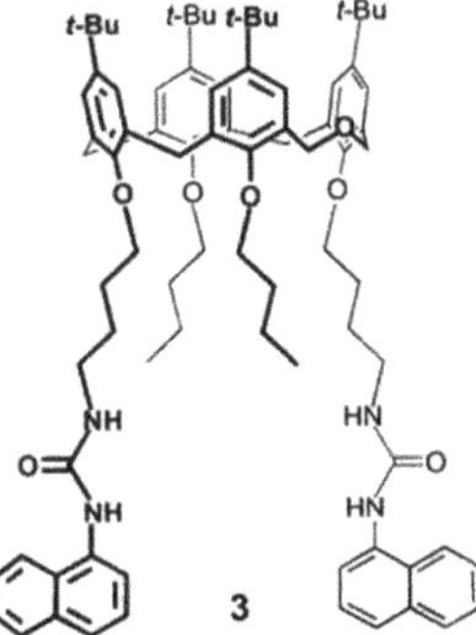

Scheme 2. Chemical structure of compound **3**.

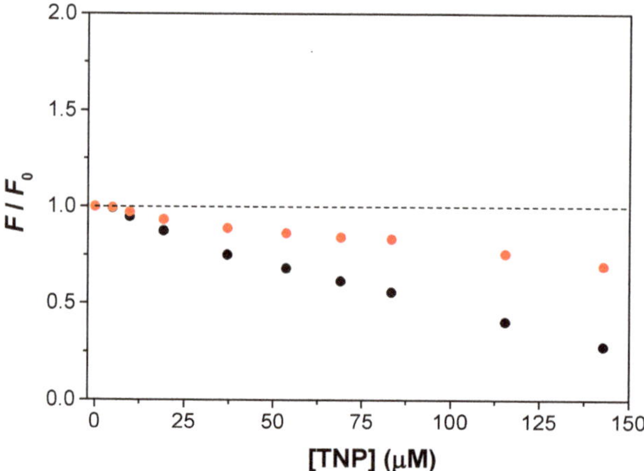

Figure 7. Relative fluorescence intensity (black circles) of Napht urea **3** (10 μM) upon addition of TNP (up to 20 equiv.) in CH_2Cl_2. F_0 is the intensity in the absence of quencher. Also shown are the intensities corrected for the inner filter effect (red circles). $\lambda_{ex} = 300$ nm.

Table 3. Two- and three-exponential analysis of fluorescence decays of **3** with TNP in CH_2Cl_2 at 25 °C.

	τ_1/ns (%)	τ_2/ns (%)	τ_3/ns (%)	$\bar{\tau}$/ns
3 *	0.26 (10)	2.2 (12)	11.2 (78)	9.0
3 + 4 equiv. TNP	0.25 (10)	2.0 (12)	10.9 (78)	8.7
3 + 10 equiv. TNP	0.24 (10)	2.1 (12)	10.8 (77)	8.6

* $\lambda_{ex} = 300$ nm.

Stern-Volmer plots of uncorrected and corrected fluorescence intensity data are shown in Figure 8. It is observed that the uncorrected data deviate from linearity for the higher concentrations, displaying a pronounced upward curvature. Note that, for the sake of clarity in the comparison with corrected data, the last four points are omitted from Figure 8. The slopes are the association constants in μM^{-1}. It is thus seen that the uncorrected data yield an apparent association constant of 9400 M^{-1}, whereas the corrected data yield an association constant of 3000 M^{-1} for a 1:1 complex.

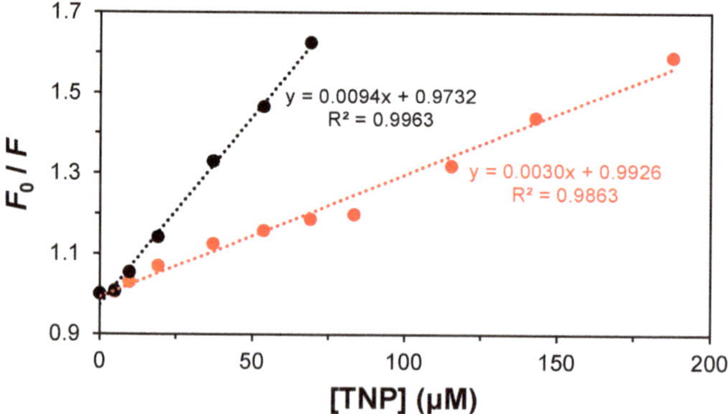

Figure 8. Linear regression of F_0/F versus [TNP] (black circles) of Napht urea **3** (10 μM) upon addition of TNP (up to 20 equiv.) in CH_2Cl_2. F_0 is the intensity in the absence of quencher. Also shown is the same linear regression for the intensities corrected for the inner filter effect (red circles). $\lambda_{ex} = 300$ nm.

A sequential approach to the use of fluorescence in quenching studies, namely with NACs, is depicted in Scheme 3, stressing the need to check for potential inner filter effects, both in excitation (absorption by the quencher at the excitation wavelength) and emission (absorption by the quencher at the emission wavelength). We concentrate here on the excitation inner filter effect. In some cases, a change in the excitation wavelength suffices to remove the interference. However, when extensive spectral overlap of the absorption spectra of the fluorescent compound and quencher (NAC) exists, this may not be possible. The use of cells with a small optical path can also work, if enough fluorescence intensity is still obtained, and if the quencher absorption is weak. In any case, a correction should still be applied, as is done here, even if only to confirm that the inner filter effect is indeed negligible. Note that the correction depends on the emitting area sampled by the detector. In many systems, as it happens in this work, only the central part of the cell matters. However, there are correction formulas in the literature that do not apply to this case [18], so users should check the situation for the fluorimeter used, in order not to apply inadequate correction formulas.

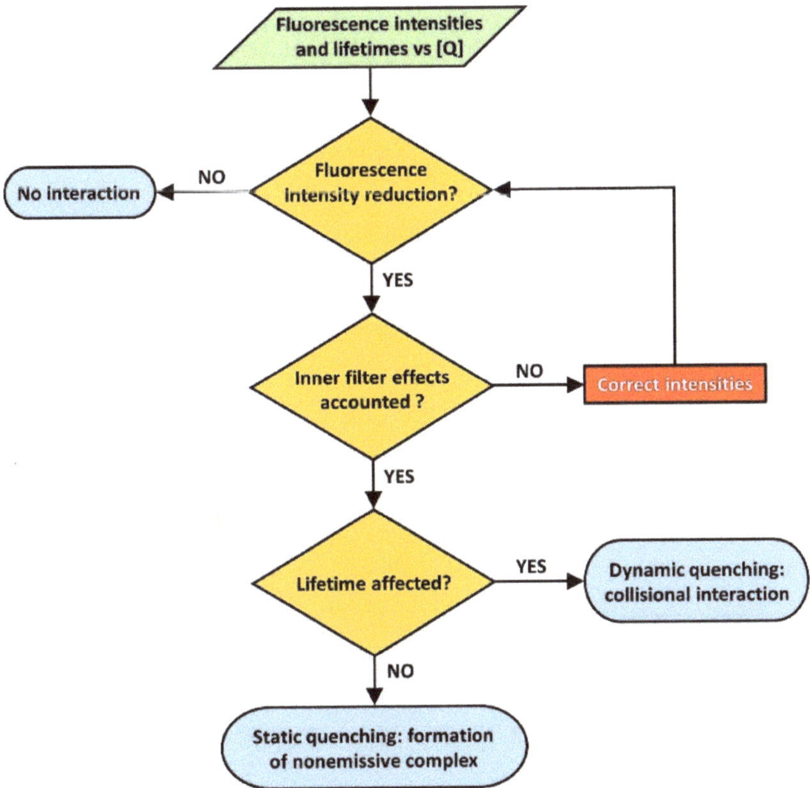

Scheme 3. Simplified sequence for the analysis of fluorescence quenching, highlighting the role of potential inner filter corrections. To simplify, only the two main types of quenching are considered, thus discarding situations of mixed quenching.

2.1.2. NMR Studies

Although unlikely, it is conceivable, in principle, that the host and guest can associate without either the electronic absorption (of both host and guest) or the fluorescence (of the host in this case) being altered, if chromophores and fluorophores are not directly involved in the binding or indirectly affected by it. For this reason, the interaction between ureas **1** and **2** and guests TNP, NB and DNB was also studied by NMR spectroscopy, in an effort to add further insights into the possible binding of the NACs. Proton NMR titrations were performed by adding increasing amounts of the NACs (up to 50 equiv.) to CDCl$_3$ solutions of **1** and **2**. Due to the C_s-symmetry of the macrocycles, two sets of naphthyl/pyrenyl aromatic protons are present in the spectra, resulting in extensive peak overlapping and, consequently, greater difficulty to follow them during the titrations. However, it is possible to observe small upfield shift variations ($\Delta\delta \leq 0.06$ and 0.15 ppm, respectively, for **1** and **2**) upon addition of TNP to the hosts (see, for example, Figure S16 for **1** + TNP). This shielding effect may suggest weak interactions between the fluorophore moieties and the guest. No variations were seen in the NMR spectra of both ureas throughout the titrations with DNB and NB (see Figure S17 for **1** + DNB).

2.2. Computational Studies

2.2.1. Semiempirical Calculations

To further investigate any possible complexation between hexahomotrioxacalix[3]arene ureas **1** and **2** and the NACs, they were subjected to computational analysis. The calixarenes, NAC guests and their complexes were built in Spartan 20 [26] and geometry optimised using molecular mechanics (MMFF) before gas phase semiempirical calculations (PM6) were undertaken using our previously published methods [27,28]. The resulting structures (Figures S18 and S19) indicate the possibility of π–π interactions in some cases, so the thermodynamics of the systems were calculated.

To investigate the stabilities of these complexes, $\Delta G_{\text{(binding)}}$ was calculated from (2):

$$\Delta G_{\text{(binding)}} = \Delta G_{\text{(complex)}} - \left[\Delta G_{\text{(host)}} + \Delta G_{\text{(binding)}}\right] \qquad (2)$$

An initial investigation, in which the NACs were bound by the macrocycles, generated positive values suggesting that the Gibbs energies of binding were unfavourable. Subsequent calculations, in which the NACs interacted with the macrocycle aromatic substituents, resulted in very negative values (Table S1). While the data support NAC binding, the binding energies are, as is not uncommon with semiempirical calculations, unrealistically large. Consequently, higher level DFT calculations were undertaken.

2.2.2. QM Calculations

The binding of the different guests with naphthyl urea **1** was also studied by a QM/MM approach. The optimisation of the host shows energy minima structures, where the two naphthyl urea arms oriented on the same side of the macrocyclic ring (pair P) interact through π-stacking and hydrogen bonds, while the inverted arm (M) is isolated. In solution, the guests may interact with the inverted arm (M) or with the pair of arms (P). The computed binding energies obtained (Table 4) indicated that the guests form very labile complexes with naphthyl urea **1**, as shown by the positive or almost zero values, except for TNP for which the Gibbs energy suggests some binding.

Table 4. QM/MM computed $\Delta G_{\text{(binding)}}$ (kJ·mol^{-1}) for complexes of **1** with NAC guests.

	1	
	M	**P**
NT	9.2	3.3
DNT	-0.4	3.3
TNT	2.1	1.7
TNP	-1.7	-13.0
NB	12.6	8.8
DNB	9.6	2.5

M denotes inverted urea arm. P denotes pair of urea arms on the same side of the macrocycle.

Further refinement of the structures using DFT methods and analysis of interaction modes between naphthyl urea **1** and a given guest (i.e., DNB) were performed using B3LYP/6-31G(d,p) optimisations in the gas phase, resulting in twenty different initial positions of the guest, leading to fourteen different minimised conformations. Optimised structures and energies of free **1** and DNB $\subset$ **1** are given in Figures S20, S21 and Table S2.

The most stable conformation of free **1** possesses intramolecular H-bonds between the two urea groups oriented on the same side of the macrocyclic ring, and it is more stable than the two other minima by about 37–42 kJ·mol^{-1} (Figure S20). The QM optimisations of the DNB $\subset$ **1** display a great variety of structures and interaction modes between the calixarene and the guest (Figure S21). All the conformations are within an energy range of 50 kJ·mol^{-1} and show typical host···guest interactions that can be either (i) H-bonding between the oxygen atoms of one or two nitro groups of DNB and the hydrogen atoms from one or two urea groups (see, for example, conformations 10, 8, 7 or 6); (ii) π···π interactions between the aromatic phenyl moiety of DNB and the naphthyl group (conformations 12 or 13); (iii) inclusion of the DNB in the calixarene upper rim (conformations 3 or 4); (iv) CH ··· π interactions between the aromatic ring of DNB and the urea oxygen (conformations 13 or 14). Complexation energies were calculated taking into consideration the most stable conformation of **1** (conformation 3 in Figure S20), which explains the energy differences obtained for interactions of similar nature, as for example the conformations 10 and 6 which both display a NH···O$_{NO2}$ interaction but differ in the conformation of the calixarene and thus conducted to ΔE of -51.9 vs $+3.4$ kJ·mol^{-1}. These QM calculations clearly show the large variety of loose pairs that can be formed between **1** and DBN molecules.

2.2.3. MD Simulations

MD simulations were also performed in order to better understand the structure and dynamics of the interactions between naphthyl urea **1** and the guests in solution. Systems containing one host and either TNP, DNB or NT guests were simulated during 50 ns in dichloromethane and acetonitrile solvent. Figure 9 shows initial and final configurations of urea **1** with 20 DNB molecules in both solvents.

In solution, **1** always displays intramolecular H-bonds between the two urea arms oriented on the same side of the macrocyclic ring. Meanwhile, the inverted naphthyl urea arm is very flexible and has multiple conformations over time. In this case, as with the two other guests, the DNB molecules are dispersed in the simulation box and display, during the simulation, close contact with the calixarene. In both solvents, free guests remain either isolated or involved in small loose pairs formed via π···π or CH···π interactions. TNP, DNB and NT also interact with the calixarene during the simulation, as illustrated in Figure 10 by snapshots of typical host···guest aggregates. NACs display close contact with the naphthyl groups via π···π bonds, thus forming dimers or trimers with the naphthyl moiety.

In all the simulated systems, the global calixarene···guest interaction energies were evaluated as a function of the nature of both guest and solvent (Table 5). The influence of guest is the same in acetonitrile and dichloromethane. Total interaction energies obtained from three independent MD runs show the same trends with TNP interacting more than 10 times more strongly than NT and 2–3 times than DNB. These variations correlate with the number of urea arm···guest interactions observed during the simulations, which follow the order TNP > DNB > NT in both solutions (Table 5). In acetonitrile, the interactions are quite weak as the average number of urea arm···guest interactions is always below 1 and the lifetimes of the species are around 12–35 ps. Conversely, in dichloromethane, the higher interaction energies are linked to more frequent urea arm···guest contacts with longer lifetimes (up to 110 ps). On average, for all the guests and solvents, MD simulations show the lability and slight interactions between the calixarene and the NACs in direct connection with the experimental results.

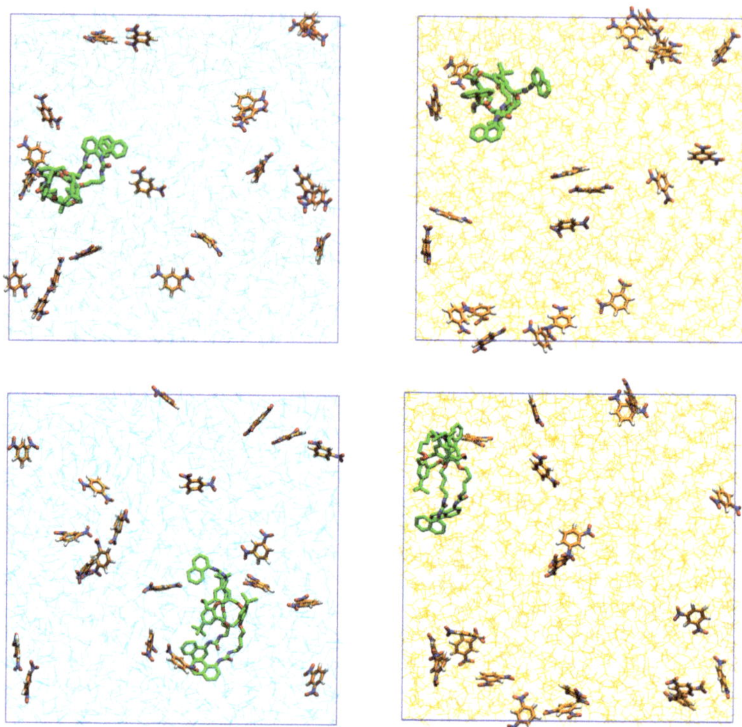

Figure 9. Initial (**top**) and final (**bottom**) snapshots of a box with one calixarene and 20 DNB molecules in acetonitrile (left, in cyan) and in dichloromethane (right, in yellow).

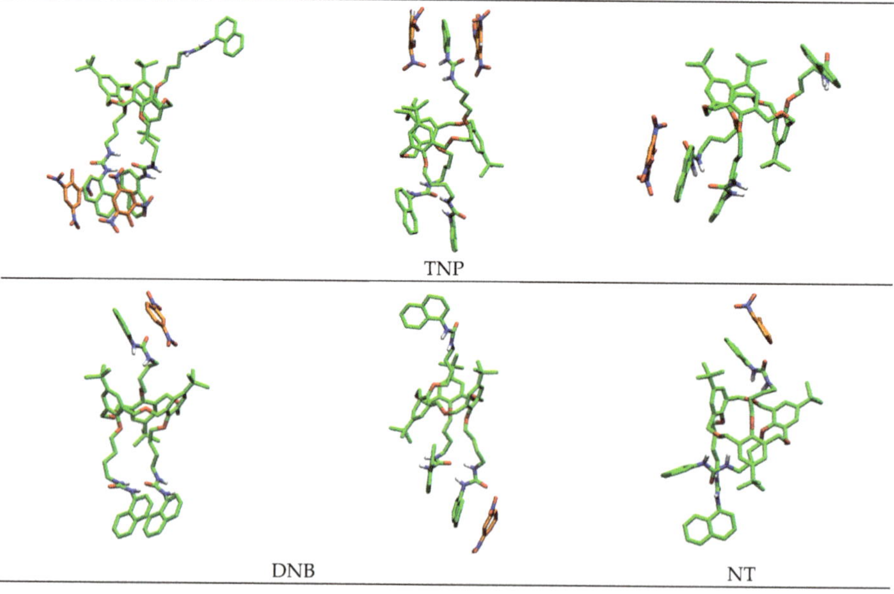

Figure 10. Typical calixarene ⋯ guest aggregates observed during the simulations (guest in orange).

Table 5. Host···guest interaction energies and fluctuations (in $kJ \cdot mol^{-1}$), number of urea arm···guest interaction (N) per arm and fluctuations (arm 1/arm 2/arm 3) and average lifetimes (in ps) [a].

		TNP	DNB	NT
MeCN	E	-110 ± 75	-29 ± 35	-11 ± 15
		-123 ± 52	-31 ± 31	-13 ± 16
		-123 ± 52	-22 ± 26	-12 ± 16
	N	$0.54/0.34/0.82 \pm 0.68$ (24–35)	$0.30/0.26/0.30 \pm 0.52$ (22–35)	$0.21/0.18/0.13 \pm 0.42$ (14–15)
		$0.74/0.41/0.50 \pm 0.58$ (20–34)	$0.32/0.23/0.44 \pm 0.47$ (20–35)	$0.17/0.16/0.17 \pm 0.43$ (14–15)
		$0.30/0.37/0.97 \pm 0.53$ (24–28)	$0.24/0.27/0.17 \pm 0.51$ (25–30)	$0.19/0.17/0.17 \pm 0.41$ (12–15)
CH_2Cl_2	E	-246 ± 97	-86 ± 50	-17 ± 20
		-336 ± 60	-59 ± 39	-18 ± 19
		-247 ± 71	-93 ± 61	-35 ± 29
	N	$1.98/1.53/1.38 \pm 0.68$ (50–90)	$0.58/0.33/0.83 \pm 0.50$ (21–37)	$0.27/0.21/0.22 \pm 0.42$ (12–15)
		$1.30/1.69/1.79 \pm 0.60$ (50–110)	$0.33/0.51/0.23 \pm 0.52$ (15–30)	$0.20/0.19/0.26 \pm 0.40$ (14–16)
		$1.39/1.10/1.56 \pm 0.55$ (40–85)	$0.43/0.51/0.54 \pm 0.45$ (28–40)	$0.32/0.21/0.27 \pm 0.42$ (14–16)

[a] Napht urea **1**···**20** molecule guests in MeCN and CH_2Cl_2 (3 separate runs).

These conclusions are also consistent with the QM analysis (see above) and with the value of the calculated electrostatic potential maps (Figure S22) that indicate a better affinity between a naphthyl group, which has roughly a -50 $kJ \cdot mol^{-1}$ potential, and TNP with a highly positive potential ($+165$ $kJ \cdot mol^{-1}$) compared to DNB ($+95$ $kJ \cdot mol^{-1}$) and NT ($+12$ $kJ \cdot mol^{-1}$), thus explaining the large differences observed during the MD simulations.

3. Materials and Methods

The naphthyl (**1**) and pyrenyl (**2**) ureido-hexahomotrioxacalix[3]arene compounds studied in this work were previously synthesised [21], as well as the naphthyl urea dihomooxacalix[4]arene **3** [20].

3.1. UV-Vis Absorption and Fluorescence Studies

Absorption and fluorescence studies were carried out using a Shimadzu UV-3101PC UV-Vis-NIR spectrophotometer and a Fluorolog F112A fluorimeter in right-angle configuration, respectively. The studies were made in CH_2Cl_2 and MeCN at 25 °C. The absorption spectra were recorded between 225 and 400 nm and the emission ones between 320 and 550 nm and using quartz cells with an optical path length of 1 cm. The excitation wavelengths used were at the maximum absorption of the calixarenes, using a right-angle geometry. The titrations were performed cell by cell with different concentrations of nitroaromatic compounds up to 30 equiv. and with a constant concentration of the receptors (10–20 µM). Emission spectra were corrected for the spectral response of the optics and the photomultiplier. In addition, the emission spectra were further corrected for the internal filter effect. The spectral changes were interpreted using the HypSpec 2014 program [29]. Time-resolved fluorescence intensity decays were obtained using the single-photon timing method with laser excitation and microchannel plate detection, with the set-up already described [30]. The excitation wavelength used was at the maximum absorption of the calixarene and the emission wavelength at the maximum emission, using a right-angle geometry. Decay data analysis with a sum of exponentials was achieved by means of a Microsoft Excel spreadsheet specially designed for lifetime analysis that considers the convolution with the IRF.

3.2. ^{1}H NMR Studies

Several aliquots (up to 50 equiv.) of the nitroaromatic compounds (TNP, NB and DNB) in $CDCl_3$ were added to $CDCl_3$ solutions of the hosts (1.25×10^{-3} M) directly in the NMR tube. The spectra were recorded on a Bruker Avance III 500 spectrometer after each addition of the NAC's, and the temperature of the NMR probe was kept constant at 25 °C.

3.3. Computational Details

3.3.1. QM/MM Calculations

All QM/MM calculations were performed with a Gaussian 09 program [31] using an ONIOM approach [32]. The QM part was carried out at a DFT level of theory with wB97XD functional [33], which includes dispersion corrections. Atoms were described by Pople's basis set 6-31+G** [34]. The MM part was described by the Universal Force Field UFF [35]. Calculations were performed in solution and the solvent (DCM) described by a PCM [36]. Full geometry optimisations were performed, followed by a frequency calculation. Gibbs free energies were extracted from this frequency calculation.

3.3.2. DFT Calculations

Geometry optimisations were performed with the Gaussian 09 program [31] with the B3LYP density functional [37] and with the 6-31G(d,p) basis set. Starting structures for geometry optimisation were constructed by hand with Spartan [26]. All reported structures were confirmed as energy minima, with no negative eigenvalue in the Hessian matrix. Electrostatic potential maps were performed with Spartan [26]. Details concerning the molecular dynamics simulations are given in the Supporting Information.

4. Conclusions

The recognition of several nitroaromatic compounds by two fluorescent ureido-hexahomotrioxacalix[3]arenes bearing naphthyl or pyrenyl residues at the lower rim was investigated by different analytical and theoretical methods. As inferred from electronic absorption, fluorescence, NMR spectroscopy and computational studies, naphthyl urea **1** and pyrenyl urea **2**, both containing aromatic moieties, have no significant interaction with nitroaromatic compounds in both dichloromethane and acetonitrile solutions. DFT calculations performed suggest the formation of very weak and labile complexes between the calixarenes and the NACs, this being corroborated by MD simulations. On the other hand, the observed fluorescence quenching is spurious and can be entirely ascribed to an excitation inner filter effect, whereby the fluorescence decrease results from significant absorption of the excitation radiation by the NAC. This is in stark contrast with the conclusions from some previous reports on NAC sensing using the fluorescence of calixarenes bearing long wavelength absorbing aromatic moieties. A naphthyl urea dihomooxacalix[4]arene (**3**) investigated is also subject to the inner filter effect but forms a stable complex with TNP; however, the equilibrium association constant is largely overestimated (9400 M^{-1} vs 3000 M^{-1}) if no correction to the inner filter effect is applied, again stressing the importance of taking into account this effect.

Supplementary Materials: The following supporting information can be downloaded at https://www.mdpi.com/article/10.3390/molecules28073052/s1, UV-Vis absorption spectra and fluorescence intensity plots; NMR titration spectra; semiempirical calculations; details of molecular dynamics simulations; DFT calculations. References [38–44] cited in the Supplementary material.

Author Contributions: A.S.M.: steady-state fluorescence and absorption experiments, data acquisition and analysis, editing; P.M.M.: supervision, NMR data acquisition, analysis and interpretation, writing and editing; J.R.A.: NMR data analysis and interpretation; M.N.B.-S.: conceptualisation, supervision, photophysics data analysis, interpretation and writing; P.J.C.: MM and semiempirical calculations, analysis and writing; R.S.: DFT calculations and MD simulations, analysis and writing; C.G.: QM/MM calculations, analysis and writing. All authors contributed to the discussion and final form of the paper. All authors have read and agreed to the published version of the manuscript.

Funding: The authors thank Fundação para a Ciência e a Tecnologia, Projects UIDB/00100/2020 and UIDB/04565/2020. A. S. Miranda thanks a PhD Grant ref. SFRH/BD/129323/2017 and COVID/BD/152147/2022.

Institutional Review Board Statement: Not applicable.

Informed Consent Statement: Not applicable.

Data Availability Statement: Not applicable.

Acknowledgments: The authors thank Alexander Fedorov for performing the fluorescence lifetime measurements.

Conflicts of Interest: The authors declare no conflict of interest.

Sample Availability: Samples of the compounds are not available from the authors.

References

1. Sun, X.; Wang, Y.; Lei, Y. Fluorescence based explosive detection: From mechanisms to sensory materials. *Chem. Soc. Rev.* **2015**, *44*, 8019–8061. [CrossRef] [PubMed]
2. Martelo, L.M.; Marques, L.F.; Burrows, H.D.; Berberan-Santos, M.N. Explosive detection: From sensing to response. In *Fluorescence in Industry*; Pedras, B., Ed.; Springer Series on Fluorescence; Springer: Cham, Switzerland, 2019; Volume 18.
3. Moore, D.S. Recent advances in trace explosives detection instrumentation. *Sens. Imaging.* **2007**, *8*, 9–38. [CrossRef]
4. Meaney, M.S.; McGuffin, V.L. Luminescence-based methods for sensing and detection of explosives. *Anal. Bioanal. Chem.* **2008**, *391*, 2557–2576. [CrossRef] [PubMed]
5. Rasheed, T.; Nabeel, F.; Rizwan, K.; Bilal, M.; Hussain, T. Conjugated supramolecular architectures as state-of-the-art materials in detection and remedial measures of nitro based compounds: A review. *Trends Anal. Chem.* **2020**, *129*, 115958. [CrossRef]
6. Kumar, R.; Sharma, A.; Singh, H.; Suating, P.; Kim, H.S.; Sunwoo, K.; Shim, I.; Gibb, B.C.; Kim, J.S. Revisiting fluorescent calixarenes: From molecular sensors to smart materials. *Chem. Rev.* **2019**, *119*, 9657–9721. [CrossRef]
7. Desai, V.; Panchal, M.; Dey, S.; Panjwani, F.; Jain, V.K. Recent advancements for the recognition of nitroaromatic explosives using calixarene based fluorescent probes. *J. Fluoresc.* **2022**, *32*, 67–79. [CrossRef] [PubMed]
8. Gutsche, C.D. *Calixarenes, An Introduction; Monographs in Supramolecular Chemistry*; The Royal Society of Chemistry: Cambridge, UK, 2008.
9. Neri, P.; Sessler, J.L.; Wang, M.-X. (Eds.) *Calixarenes and Beyond*; Springer International Publishing: Cham, Switzerland, 2016.
10. Lee, Y.H.; Liu, H.; Lee, J.Y.; Kim, S.H.; Kim, S.K.; Sessler, J.L.; Kim, Y.; Kim, J.S. Dipyrenylcalix[4]arene—A fluorescence-based chemosensor for trinitroaromatic explosives. *Chem. Eur. J.* **2010**, *16*, 5895–5901. [CrossRef] [PubMed]
11. Zhan, J.; Zhu, X.; Fang, F.; Miao, F.; Tian, D.; Li, H. Sensitive fluorescence sensor for nitroaniline isomers based on calix[4]arene bearing naphthyl groups. *Tetrahedron* **2012**, *68*, 5579–5582. [CrossRef]
12. Zhang, F.; Luo, L.; Sun, Y.; Miao, F.; Bi, J.; Tan, S.; Tian, D.; Li, H. Synthesis of a novel fluorescent anthryl calix[4]arene as picrid acid sensor. *Tetrahedron* **2013**, *69*, 9886–9889. [CrossRef]
13. Cao, X.; Luo, L.; Zhang, F.; Miao, F.; Tian, D.; Li, H. Synthesis of a deep cavity calix[4]arene by fourfold Sonogashira cross-coupling reaction and selective fluorescent recognition toward p-nitrophenol. *Tetrahedron Lett.* **2014**, *55*, 2029–2032. [CrossRef]
14. Boonkitpatarakul, K.; Yodta, Y.; Niamnont, N.; Sukwattanasinitt, M. Fluorescent phenylethynylene calix[4]arenes for sensing TNT in aqueous media and vapor phase. *RSC Adv.* **2015**, *5*, 33306–33311. [CrossRef]
15. Bandela, A.K.; Bandaru, S.; Rao, C.P. A fluorescent 1,3-diaminonaphthalimide conjugate of calix[4]arene for sensitive and selective detection of trinitrophenol: Spectroscopy, microscopy, and computational studies, and its applicability using cellulose strips. *Chem. Eur. J.* **2015**, *21*, 13364–13374. [CrossRef] [PubMed]
16. Dinda, S.K.; Hussain, M.A.; Upadhyay, A.; Rao, C.P. Supramolecular sensing of 2,4,6-trinitrophenol by a tetrapyrenyl conjugate of calix[4]arene: Applicability in solution, in solid state, and on the strips of cellulose and silica gel and the image processing by a cellular phone. *ACS Omega* **2019**, *4*, 17060–17071. [CrossRef] [PubMed]
17. Prata, J.V.; Costa, A.I.; Teixeira, C.M. A solid-state fluorescence sensor for nitroaromatics and nitroanilines based on a conjugated calix[4]arene polymer. *J. Fluoresc.* **2020**, *30*, 41–50. [CrossRef]
18. Barata, P.; Prata, J.V. Fluorescent calix[4]arene-carbazole-containing polymers as sensors for nitroaromatic explosives. *Chemosensors* **2020**, *8*, 128. [CrossRef]
19. Miranda, A.S.; Martelo, L.M.; Fedorov, A.A.; Berberan-Santos, M.N.; Marcos, P.M. Fluorescence properties of p-tert-butyldihomooxacalix[4]arene derivatives and the effect of anion complexation. *New J. Chem.* **2017**, *41*, 5967–5973. [CrossRef]
20. Miranda, A.S.; Marcos, P.M.; Ascenso, J.R.; Berberan-Santos, M.N.; Schurhammer, R.; Hickey, N.; Geremia, S. Dihomooxacalix[4]arene-based fluorescent receptors for anion and organic ion pair recognition. *Molecules* **2020**, *25*, 4708. [CrossRef]
21. Miranda, A.S.; Marcos, P.M.; Ascenso, J.R.; Berberan-Santos, M.N.; Menezes, F. Anion binding by fluorescent ureido-hexahomotrioxacalix[3]arene receptors: An NMR, absorption and emission spectroscopic study. *Molecules* **2022**, *27*, 3247. [CrossRef]
22. Cottet, K.; Marcos, P.M.; Cragg, P.J. Fifty years of oxacalix[3]arenes: A review. *Beilstein J. Org. Chem.* **2012**, *8*, 201–226. [CrossRef]
23. Marcos, P.M. Functionalization and properties of homooxacalixarenes. In *Calixarenes and Beyond*; Neri, P., Sessler, J.L., Wang, M.-X., Eds.; Springer International Publishing: Cham, Switzerland, 2016; pp. 445–466.
24. Wu, C.; Zhao, J.-L.; Jiang, X.-K.; Ni, X.-L.; Zeng, X.; Redshaw, C.; Yamato, T. Click-modified hexahomotrioxacalix[3]arenes as fluorometric and colorimetric dual-modal chemosensors for 2,4,6-trinitrophenol. *Anal. Chim. Acta* **2016**, *936*, 216–221. [CrossRef]
25. Sarova, G.; Berberan-Santos, M.N. Stable charge-transfer complexes versus contact complexes. Application to the interaction of fullerenes with aromatic hydrocarbons. *J. Phys. Chem. B* **2004**, *108*, 17261.

26. *Spartan 20 software*, Version 1.1.3; Wavefunction Inc.: Irvine, CA, USA.

27. Sheehan, R.; Cragg, P.J. Supramolecular chemistry in silico. *Supramol. Chem.* **2008**, *20*, 443–451. [CrossRef]

28. Sambrook, M.R.; Vincent, J.C.; Ede, J.A.; Gass, I.A.; Cragg, P.J. Experimental and computational study of the inclusion complexes of β-cyclodextrin with the chemical warfare agent soman (GD) and commonly used simulants. *RSC Adv.* **2017**, *7*, 38069–38076. [CrossRef]

29. Gans, P.; Sabatini, A.; Vacca, A. Investigation of equilibria in solution. Determination of equilibrium constants with the HYPERQUAD suite of programs. *Talanta* **1996**, *43*, 1739–1753. [CrossRef]

30. Menezes, F.; Fedorov, A.; Baleizão, C.; Valeur, B.; Berberan-Santos, M.N. Methods for the analysis of complex fluorescence decays: Sum of Becquerel functions versus sum of exponentials. *Methods Appl. Fluoresc.* **2013**, *1*, 015002. [CrossRef]

31. Frisch, M.J.; Trucks, G.W.; Schlegel, H.B.; Scuseria, G.E.; Robb, M.A.; Cheeseman, J.R.; Scalmani, G.; Barone, V.; Petersson, G.A.; Nakatsuji, H.; et al. Gaussian 09, Rev. D.01. Wallingford, CT, USA. 2016. Available online: https://gaussian.com (accessed on 2 March 2016).

32. Dapprich, S.; Komáromi, I.; Byun, K.S.; Morokuma, K.; Frisch, M.J. A New ONIOM implementation in Gaussian 98. Part 1. The calculation of energies, gradients and vibrational frequencies and electric field derivatives. *J. Mol. Struct. THEOCHEM* **1999**, *461–462*, 1–21. [CrossRef]

33. Chai, J.-D.; Head-Gordon, M. Long-range corrected hybrid density functionals with damped atom-atom dispersion corrections. *Phys. Chem. Chem. Phys.* **2008**, *10*, 6615–6620. [CrossRef]

34. Ditchfield, R.; Hehre, W.J.; Pople, J.A. Self-consistent molecular orbital methods. IX. Extended Gaussian-type basis for molecular-orbital studies of organic molecules. *J. Chem. Phys.* **1971**, *54*, 724. [CrossRef]

35. Rappé, A.K.; Casewit, C.J.; Colwell, K.S.; Goddard III, W.A.; Skiff, W.M. UFF, a full periodic-table force-field for molecular mechanics and molecular-dynamics simulations. *J. Am. Chem. Soc.* **1992**, *114*, 10024–10035. [CrossRef]

36. Miertuš, S.; Scrocco, E.; Tomasi, J. Electrostatic interaction of a solute with a continuum. A direct utilization of ab initio molecular potentials for the prevision of solvent effects. *Chem. Phys.* **1981**, *55*, 117–129. [CrossRef]

37. Becke, A.D. Density functional thermochemistry. III. The role of exact exchange. *J. Chem. Phys.* **1993**, *98*, 5648–5652. [CrossRef]

38. Case, D.A.; Ben-Shalom, I.Y.; Brozell, S.R.; Cerutti, D.S.; Cheatham, E.; Cruzeiro, V.W.D.; Darden, T.A.; Duke, R.E.; Ghoreishi, D.; Gilson, M.K.; et al. AMBER 18. University of California: San Francisco, CA, USA, 2019.

39. Wang, J.; Wolf, R.M.; Caldwell, J.W.; Kollman, P.A. Development and testing of a general amber force field. *J. Comput. Chem.* **2004**, *25*, 1157–1174. [CrossRef] [PubMed]

40. Bayly, C.I.; Cieplak, P.; Cornell, W.D.; Kollman, P.A. A well-behaved electrostatic potential based method using charge restraints for deriving atomic charges: The RESP model. *J. Phys. Chem.* **1993**, *97*, 10269–10280. [CrossRef]

41. Fox, T.; Kollman, P.A. Application of the RESP methodology in the parametrization of organic solvents. *J. Phys. Chem. B* **1998**, *102*, 8070–8079. [CrossRef]

42. Jorgensen, W.L.; Briggs, J.M. Monte Carlo simulations of liquid acetonitrile with a three-site model. *Mol. Phys.* **1988**, *63*, 547–558. [CrossRef]

43. Allen, M.P.; Tildesley, D.J. Computer simulation of liquids. Clarendon Press: Oxford, UK, 1987.

44. Humphrey, W.; Dalke, A.; Schulten, K. VMD: Visual molecular dynamics. *J. Mol. Graph.* **1996**, *14*, 33–38. [CrossRef]

Article

A Combined Solution and Solid-State Study on the Tautomerism of an Azocalix[4]arene Chromoionophore

Laura Baldini *, Davide Balestri, Luciano Marchiò and Alessandro Casnati

Department of Chemistry, Life Sciences and Environmental Sustainability, University of Parma, Parco Area delle Scienze 17/a, 43124 Parma, Italy; davide.balestri@unipr.it (D.B.); luciano.marchio@unipr.it (L.M.); alessandro.casnati@unipr.it (A.C.)
* Correspondence: laura.baldini@unipr.it

Abstract: Azocalixarenes functionalized with cation binding sites are popular chromoionophores due to the ease of synthesis and the large complexation-induced shifts of their absorption band that originate from an azo-phenol–quinone-hydrazone tautomerism. Despite their extensive use, however, a thorough investigation of the structure of their metal complexes has not been reported. We describe herein the synthesis of a new azocalixarene ligand (**2**) and the study of its complexation properties with the Ca^{2+} cation. Through a combination of solution (^{1}H NMR and UV-vis spectroscopies) and solid-state (X-ray diffractometry) techniques, we demonstrate that metal complexation induces a shift of the tautomeric equilibration towards the quinone-hydrazone form, while deprotonation of the complex results in the reversion to the azo-phenol tautomer.

Keywords: chromoionophores; calix[4]arenes; azo-phenol–quinone-hydrazone tautomerism; Ca^{2+} complexation

Citation: Baldini, L.; Balestri, D.; Marchiò, L.; Casnati, A. A Combined Solution and Solid-State Study on the Tautomerism of an Azocalix[4]arene Chromoionophore. *Molecules* **2023**, *28*, 4704. https://doi.org/10.3390/molecules28124704

Academic Editor: Paula M. Marcos

Received: 12 May 2023
Revised: 5 June 2023
Accepted: 6 June 2023
Published: 12 June 2023

1. Introduction

Chromoionophores are synthetic molecular receptors that combine in their structure an ionophoric site for the complexation of an ion analyte and a chromogenic moiety that changes color in response to the recognition event. Crucial for the correct functioning of chromoionophore-based sensors is the connection between the sensing and reporting units. In the case of chromoionophores for the optical detection of metal cations, a fast and strong response can generally be achieved if a portion of the chromogenic moiety takes part in the coordination site. In this way, upon complexation, the excited state of the chromophore can be stabilized or destabilized more strongly than the ground state, resulting in a bathochromic or hypsochromic shift, respectively, of the electronic absorption band [1].

Calixarene-based chromoionophores represent one of the first and most successful applications of calixarene chemistry. Thanks to the ease of functionalization of the two rims, the calixarene scaffold can be equipped with both cation binding sites and chromophoric reporter units that undergo a color change upon metal ion complexation [2–9].

Azocalixarenes, which have the phenylazo (-N=N-Ph) group linked to the para-position of one (or more) phenol rings, are among the most studied chromogenic calixarenes. The popularity of these compounds can be ascribed to the synthetically appealing one-step diazo-coupling reaction that, pioneered in 1989 by Shinkai [10], gives good results on the calixarene phenol rings. Moreover, in azocalixarenes, the resulting 4-phenylazophenol chromophore can directly participate in cation complexation with the phenol oxygen atom as a donor site. A vast number of azocalixarene chromoionophores have been reported, with different additional binding sites for cations and different numbers and functionalizations of the 4-phenylazophenol moieties [5,7,9,11–21].

The 4-phenylazophenol moiety is well known to undergo a tautomeric equilibration between the azo-phenol and quinone-hydrazone forms (Figure 1), which depends on the substituents of the phenyl ring and on the solvent [22]. Electron-donating substituents generally stabilize the azo-phenol tautomer, while electron-withdrawing ones favor the quinone-hydrazone form. Moreover, due to the stronger hydrogen-bonding ability of the OH group of the azo-phenol compared to the NH group of the quinone-hydrazone, hydrogen-bonding acceptor solvents (such as pyridine and acetone) usually stabilize the former, while hydrogen-bonding donor solvents (such as acetic acid and chloroform) favor the latter form.

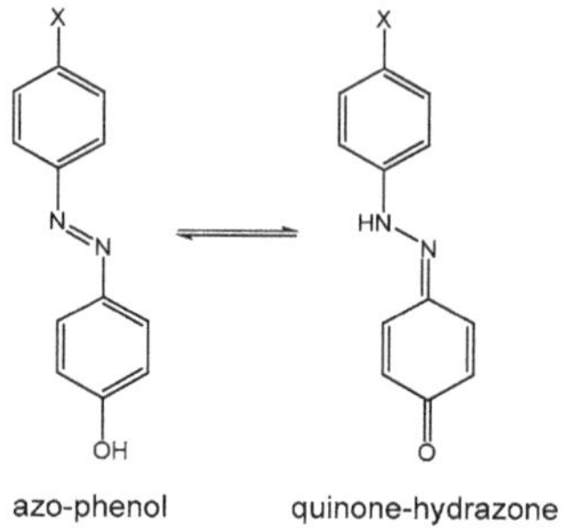

Figure 1. Azo-phenol–quinone-hydrazone tautomerism of 4-phenylazophenol.

The two tautomeric forms can be easily distinguished by UV-vis spectroscopy. The absorption band of the azo-phenol tautomer is typically displayed around 400 nm, while it is bathochromically shifted to ~480 nm for the quinone-hydrazone form [22–25].

When azocalixarenes form complexes with metal cations, large shifts of the absorption band of the azo-phenol moiety are observed, which have been attributed, in some cases, to a cation-induced stabilization of the quinone-hydrazone form [14,17,26,27] and, in others, to a metal-induced deprotonation of the ligand [18,21]. Despite their large use as chromoionophores, the characterization of the complexes is mostly achieved exclusively by UV-vis spectroscopy. In just a few papers, the complexes have also been studied by ^{1}H NMR [7,13,21], and only one crystal structure of an azocalixarene metal complex is present in the literature [20].

To fill this gap, we report herein the synthesis of a new azocalixarene chromoionophore (**2**) and the thorough characterization of its complex with the Ca^{2+} cation both in solution and in the solid state. In **2**, the binding site for the cation at the calixarene lower rim is constituted by two phenoxyacetamide groups linked to two distal phenol rings, which provide two phenolic and two carbonyl oxygen donors, and by the two phenol OH groups of the 4-nitrophenylazophenol moieties. This coordination environment has been shown in the literature to be very efficient for the complexation of hard metal cations such as Mg^{2+} [8], Ca^{2+} [6,18], Fe^{3+} [28], or trivalent lanthanide ions [28,29]. Thanks to the combined use of UV-vis and ^{1}H NMR spectroscopies and X-ray diffractometry, we have been able not only to confirm the high affinity of this ligand for the Ca^{2+} cation but also to shed light on the structure of the complex and on the tautomeric equilibration.

2. Results and Discussion

2.1. Synthesis and Characterization of **2**

Azocalixarene **2** was synthesized in only two steps, following the reaction pathway reported in Scheme 1. First, diamidocalix[4]arene **1** was obtained in good yield from the alkylation of calix[4]arene with 2-chloro-*N*,*N*-diethylacetamide, carried out according to the literature [30]. Subsequently, calixarene **1** was subjected to a diazonium coupling reaction with p-nitroaniline following a procedure adapted from [16].

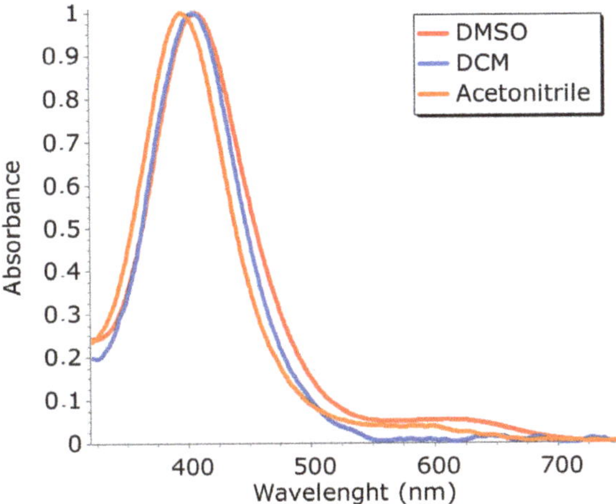

Scheme 1. Synthesis of compound **2**.

Figure 2 displays the absorption spectra of azocalixarene **2** in different solvents. Despite the presence of the electron-withdrawing nitro groups, both 4-nitrophenylazophenol moieties of **2** are present exclusively in the azo-phenol form, as indicated by the absorption band centered around 400 nm. The strong intramolecular hydrogen bonds that the phenol OH groups can donate either to the neighboring ethereal oxygen atoms or to the carbonyl C=O groups are most likely responsible for the stabilization of the azo-phenol tautomer. Similar calix[4]arene derivatives containing two 4-nitrophenylazophenol moieties in a distal position display analogous absorption spectra, confirming the shift of the tautomeric equilibration towards the azo-phenol form when this unit is embedded in the calixarene skeleton [12,21].

Figure 2. Normalized absorption spectra of **2** (1×10^{-5} M) in DMSO (red line), dichloromethane (blue line), and acetonitrile (orange line).

^{1}H NMR spectroscopy (Figure 3) shows that in acetonitrile, DMSO, and chloroform, calixarene **2** adopts the cone geometry, with the typical AX pattern of the axial and equatorial protons of the methylene bridges around 4.5 and 3.5 ppm, respectively. The signals of the hydrogen-bonded OH protons around 9.8 ppm in acetonitrile and DMSO confirm the sole presence of the azo-phenol tautomer. In chloroform, where the OH signal is missing due to chemical exchange, the same form is suggested by the invariance of all the signals with respect to those in acetonitrile and DMSO solutions.

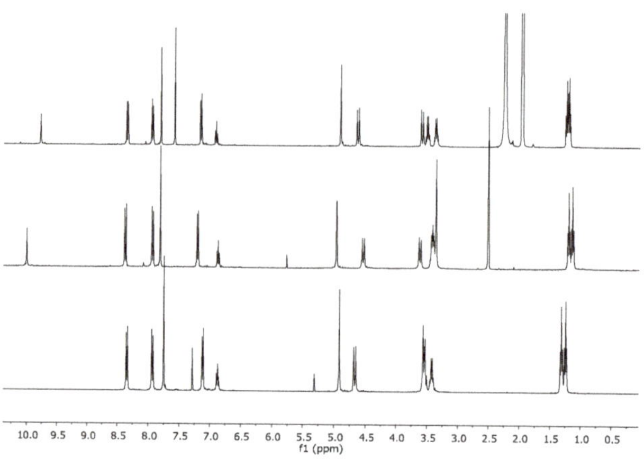

Figure 3. ^{1}H NMR spectra of **2** in different solvents. From top to bottom: CD$_3$CN-CDCl$_3$ 5:1 (*v/v*), DMSO-*d$_6$*, CDCl$_3$.

The X-ray structure analysis of crystals of **2** grown from a mixture of chloroform, methanol, and hexane (Figure 4, left) shows a flattened cone conformation of the calixarene, with the phenol rings of the phenylazophenol moieties almost parallel, having a dihedral angle of 12.5°, and with a minimum distance between the nitrophenyl rings of 3.68 Å, indicative of weak π–π interaction between the two aromatic moieties. The two phenol rings functionalized with the amide groups are tilted outward, having a dihedral angle of 67.4°. This conformation is stabilized by two strong hydrogen bonds between the phenol OH groups and the amide C=O groups (O1···O5 of approximately 2.62 Å). In the Ar-OH aromatic ring, the C-C distances vary in the 1.38–1.41 Å range, the C-O distance is 1.35 Å, and the N-N distance of the azo moiety is 1.24–1.29 Å (by taking into account the two disordered fragments, Figure S12). Taken together, these geometric parameters are suggestive of a single Ar-O bond and a double N=N bond for the azo group, pointing to the azo-phenol tautomer also in the solid state.

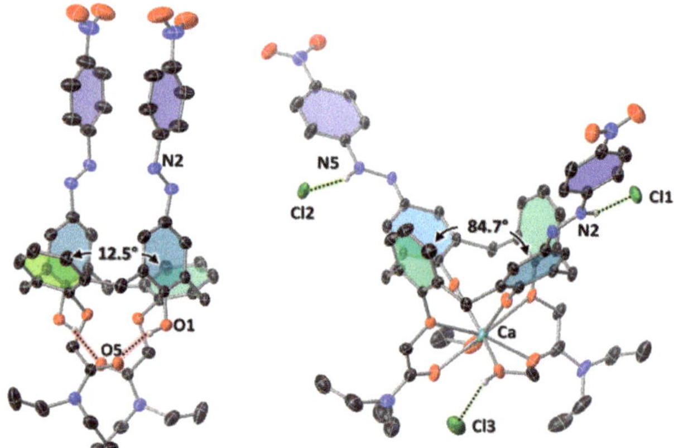

Figure 4. Molecular structure of **2** (**left**) and of **2·CaCl$_2$** (**right**). Thermal ellipsoids are depicted at the 30% probability level.

Interestingly, in the crystal packing, the nitro groups exchange O···HC interactions with the amide methyl groups of a neighboring molecule, thus giving rise to supramolecular chains that run parallel to the *b* crystallographic axis, as shown in Figure S6.

2.2. Ca^{2+} Complexation Studies

The complexation properties of chromoionophore **2** for the Ca^{2+} cation were investigated in acetonitrile solution by ^{1}H NMR titrations (Figure 5), with the aim of shedding light on the possibility of a metal ion-induced azo-phenol–quinone-hydrazone tautomerism or metal-induced deprotonation of the ligand. The addition of increasing amounts of Ca(ClO$_4$)$_2$ to a CD$_3$CN-CDCl$_3$ (6:1, *v/v*) solution of **2** resulted in the appearance of a new set of resonances consistent with the Ca^{2+} complex of ligand **2**, accompanied by the gradual disappearance of the signals of **2**. In the titration conditions, the complexation equilibrium is slow compared to the NMR timescale. After the addition of 1 equivalent of Ca^{2+}, the signals of free **2** are barely visible, while they have completely disappeared in the presence of 2 equivalents of the guest. This quasi-saturation behavior is indicative of strong binding of the cation by receptor **2**, with a binding constant that can be estimated as >10^4 M^{-1}.

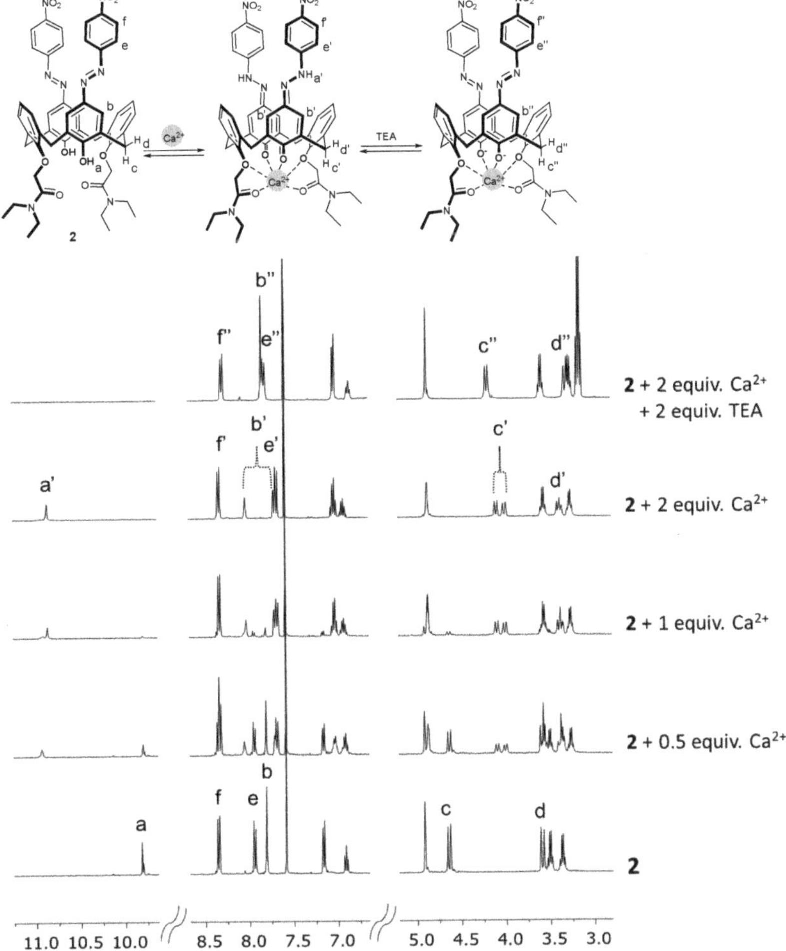

Figure 5. Partial ^{1}H NMR spectra (400 MHz, CD$_3$CN-CDCl$_3$ 6:1, *v/v*) of ligand **2** (1.4 mM) upon addition of (from bottom to top) 0, 0.5, 1, 2 equiv. of Ca(ClO$_4$)$_2$, and 2 equiv. of Ca(ClO$_4$)$_2$ + 2 equiv. of TEA. For the assignment of the peaks to protons a-f, a′–f′ and b″–f″ cfr. the calixarene structures at the top of this figure.

Besides assessing the efficiency of the complexation process, however, this experiment is particularly informative regarding the structural features of the complex and the tautomeric equilibrium. Three resonances of ligand **2** undergo significant modifications upon complex formation: (i) the signal of the phenol OH, originally seen at 9.82 ppm (a in Figure 5), is shifted to 10.87 ppm (a′); (ii) the singlet of the aromatic protons ortho to the azo group (b) is split into two signals (b′), one downfield shifted by 0.23 ppm and the other upfield shifted by 0.09 ppm; (iii) the doublet of the axial methylene bridge protons at 4.65 ppm (c) is split into two new doublets (c′), both significantly upfield shifted ($\Delta\delta$ = 0.54 and 0.64 ppm). Although similar shifts upon metal complex formation have been previously reported for analogous chromoionophores [7,13,21] and assigned to the non-symmetrical structure of the complex, with the cation occupying a lateral position of the binding cavity [7,13], in our opinion they are clearly indicative of the shift of the tautomeric equilibration towards the quinone-hydrazone moiety: (i) the hydrazone NH proton (a′), more deshielded than the phenol OH, gives rise to the singlet at 10.87; (ii) the C=N double bond is responsible for the loss of symmetry of the quinone rings and the consequent splitting of the resonances of the quinone protons (b′) that are now no longer equivalent; (iii) for the same reason, the four axial protons of the methylene bridges (c) become chemically non-equivalent in pairs and resonate as two separate doublets (c′) at a chemical shift around 4.0 ppm, which is typical for calix[4]arene diquinones [9,31].

The addition of 2 equivalents of triethylamine to the last sample of the NMR titration resulted in a color change from orange to deep purple and in modifications of the spectrum compatible with the deprotonation of the quinone-hydrazone moieties and with the consequent shift of the tautomeric equilibration back to the azo form: (i) the signal of the hydrazone NH proton (a′) disappears; (ii) the two peaks of the quinone ring at 8.04 and 7.71 ppm (b′) are replaced by a single resonance at 7.85 (b″); (iii) the signals of the methylene bridge revert to the typical AX doublets (c″ and d″). Overall, the spectrum of the deprotonated complex is characterized by the same pattern of signals as the spectrum of free **2**, with small shifts due to the presence of the complexed cation.

UV-vis spectroscopy (Figure 6) confirmed the metal complexation-induced tautomerism of the azophenol moieties: upon addition of one equivalent of $Ca(ClO_4)_2$ to a 1×10^{-5} M acetonitrile solution of **2**, the absorption band centered at 397 nm was replaced by a new band having a maximum at 494 nm, a typical value for the quinone-hydrazone tautomer [22]. Further additions of Ca^{2+} did not produce any modification of the spectrum, indicating a $K_a > 10^6$ M^{-1}. The addition of 5 equivalents of TEA to the 1:1 (**2**:Ca^{2+}) solution produced a bathochromic shift of the absorption maximum to 549 nm. Interestingly, the spectrum of a 1×10^{-5} M solution of **2** containing a large excess (~50,000 equiv.) of TEA contained two separate absorption bands, one with a maximum at 400 nm, corresponding to the azophenol moiety, and the other centered at 620 nm, due to the deprotonated azophenol group. In the absence of the cation, a weak base such as TEA is therefore able to deprotonate only one phenol group. As expected, the presence of the cation complexed at the calixarene lower rim enhances the acidity of the ligand [15,21]. The hypsochromic shift of the absorption band of the deprotonated complex (centered at 549 nm) with respect to the deprotonated azophenol in metal-free **2** (centered at 620 nm) is likely due to the presence of the cation, which hinders the negative charge delocalization from the Ar-O$^-$ to the nitro group.

Single crystals of the Ca^{2+} complex of **2** were obtained from the slow evaporation of a 1:1 solution of **2** and $CaCl_2$ in a mixture of chloroform, methanol, and hexane. In the complex, the cation is coordinated by the six oxygen atoms at the lower rim of the calixarene (the four Ar-O and the two C=O) and by two methanol molecules of crystallization (Figure 4, right). The metal geometry is distorted square antiprismatic, with Ca-O bond lengths in the 2.35–2.65 Å range (Figure S11).

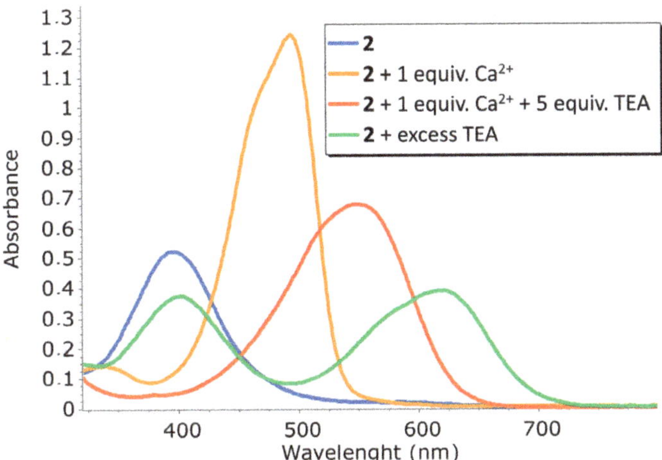

Figure 6. UV-vis spectra of ligand **2** (1×10^{-5} M in CH_3CN, blue line) and ligand **2** (1×10^{-5} M in CH_3CN) upon addition of 1 equiv. of $Ca(ClO_4)_2$ (orange line), 1 equiv. of $Ca(ClO_4)_2$ and 5 equiv. of TEA (red line), and ~50,000 equiv. of TEA (green line).

As a consequence of the cation complexation, the conformation of the calixarene scaffold undergoes considerable modification. The two aromatic rings functionalized with the azo groups, which were almost parallel in **2**, in **2·CaCl₂** are tilted outwards with a dihedral angle of 84.7°. This conformational change is required to bring the phenol oxygen atoms at a shorter distance to the cation. As a result, the other two rings, which were divergent in **2**, became almost parallel (dihedral angle of 27.1°). The overall macrocycle conformation can be described as a flattened cone that is inverted in terms of the orientation of the calixarene aromatic rings with respect to the one displayed by **2**.

The most important feature of this structure, however, is the evidence that, also in the solid state, the complexation of the cation induces the shift of the tautomeric equilibration towards the quinone-hydrazone moiety. According to the difference Fourier map, the two N atoms of the azo groups (N2 and N5) are protonated, with the hydrazone NH hydrogen atoms hydrogen bonded to the Cl⁻ anions (N2···Cl1 3.24 Å and N5···Cl2 3.28 Å, Table S3). Moreover, the bond distances within the quinone-hydrazone units are indicative of a C=O double bond (1.23 and 1.24 Å) conjugated to the two C=C double bonds of the quinone ring (in the 1.33–1.35 Å range) and to the C=N bond (1.31 and 1.32 Å) of the hydrazone group (Figure S12). The conjugation, however, does not extend to the nitrophenyl ring, which exhibits bond lengths similar to those of **2**.

Attempts to crystallize **2** in the presence of both $CaCl_2$ and triethylamine were unsuccessful. We, therefore, prepared the neutral $2^{2-} \cdot Ca^{2+}$ complex by treating a dichloromethane solution of **2** with a saturated water solution of $Ca(OH)_2$. The organic phase, whose color had changed immediately from orange to deep purple, was divided into two batches and let evaporate to grow crystals in the presence of hexane for the first batch and methanol for the second. Both solutions yielded crystals suitable for X-ray diffractometry. The two structures (**2·Ca-A** and **2·Ca-B**, respectively), albeit similarly consisting of the doubly deprotonated calixarene bound to the metal cation, present some differences.

The asymmetric unit of **2·Ca-A** comprises two calixarene units and two calcium cations that exhibit a slightly different coordination environment (Figure 7). Both cations (Ca1 and Ca2) are coordinated by the six oxygen atoms at the lower rim of the calixarene (the four Ar-O and the two C=O), but while the seventh coordination site of Ca1 is occupied by an oxygen atom of a nitro group of the second calixarene moiety, Ca2 coordinates a water molecule instead (Figure S11). The metal geometry for both cations is distorted pentagonal bipyramidal, with Ca-O bond lengths in the 2.17–2.57 Å range. The shortest

Ca-O distances are found between the metal and the deprotonated oxygen atoms of calixarene (2.17–2.25 Å range). The analysis of the bond lengths within the phenolate moiety (Figure S13) is consistent with the possible resonance forms that can accommodate the negative charge (Scheme S1). The N-N distances (in the range 1.21–1.29 Å) are consistent with a double N=N bond and support the NMR evidence that, upon deprotonation, the tautomeric equilibration reverts to the azo-phenol form.

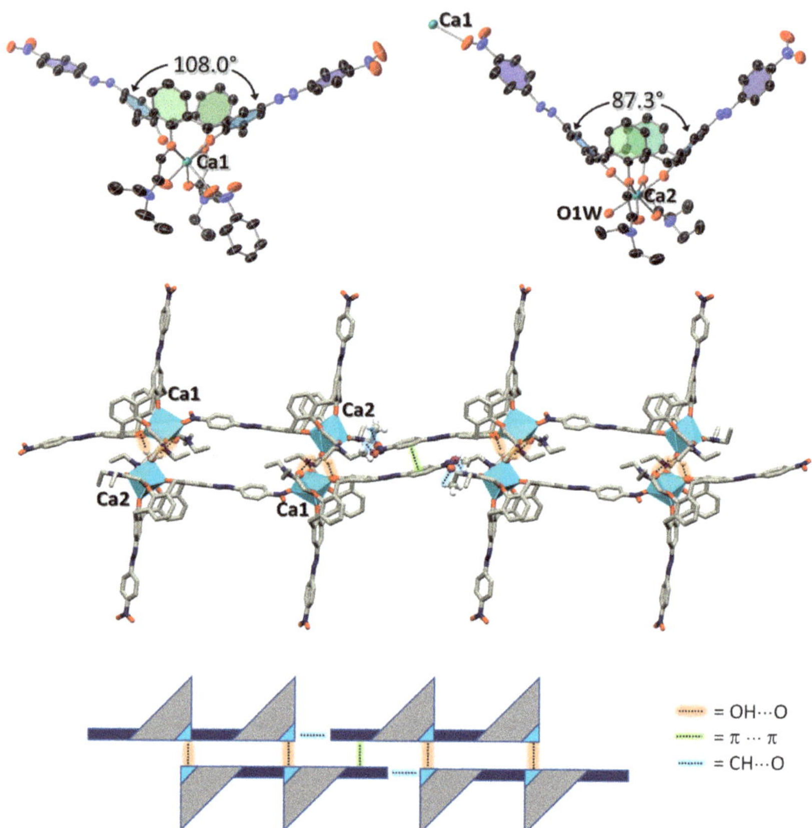

Figure 7. Top: molecular structure of the two complexes contained in the asymmetric unit for **2·Ca-A** (thermal ellipsoids are depicted at the 30% probability level). Middle: packing view and weak interactions between **2·Ca-A** dimers. Bottom: schematized view of weak supramolecular interactions.

The presence of the water molecule bound to Ca2 leads to the formation of a supramolecular tetramer sustained by two hydrogen bonds per water molecule (Figure S10). One of the nitrophenyl groups gives rise to a CH⋯O interaction with the aliphatic chains of an adjacent amide group and a π–π interaction with a symmetry-related aromatic ring. Overall, the system forms a supramolecular structure formed by two antiparallel chains (Figure 7, bottom).

In the asymmetric unit of **2·Ca-B**, two independent but very similar Ca-calixarene moieties are present, both comprising the Ca^{2+} cation bound to the lower rim of the deprotonated calixarene and without water molecules bound to the metal (Figure 8). The calixarene conformation is analogous to that found in **2·CaCl₂** and in **2·Ca-A**. The metal ion adopts a pentagonal bipyramidal geometry and is bound by six oxygen atoms of one calixarene, with the seventh position occupied by a bridging nitro group (Figure S11).

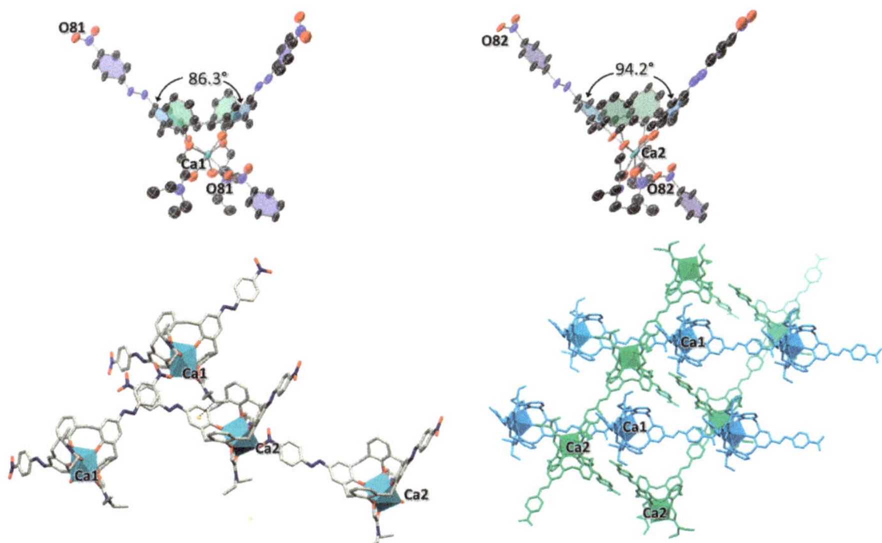

Figure 8. Top: molecular structure of the two complexes contained in the asymmetric unit for **2·Ca-B** (thermal ellipsoids are depicted at the 30% probability level). Bottom: packing view.

According to the bridging behavior of one of the nitrophenyl groups, two interpenetrated supramolecular chains represent the main packing feature. The bond lengths are consistent with the previously described resonance structure found for **2·Ca-A**.

3. Materials and Methods

3.1. General

Solvents and reagents were obtained from commercial sources and used without further purification. Analytical TLC was performed using prepared plates of silica gel (Merck 60 Merck KGaA, Darmstadt, Germany F-254 on aluminum). ^{1}H and ^{13}C NMR spectra were recorded on a Bruker Billerica, MA, USA AV400 spectrometer. All chemical shifts are reported in parts per million (ppm) using the residual peak of the deuterated solvent, whose values are referred to tetramethylsilane (TMS, $\delta_{TMS} = 0$), as internal standard. ^{13}C NMR spectra were performed with proton decoupling. Mass spectra were recorded in ESI mode on a single quadrupole instrument, SQ Detector, Waters (capillary voltage 3.7 kV, cone voltage 30–160 eV, extractor voltage 3 eV, source block temperature 80 °C, desolvation temperature 150 °C, and cone and desolvation gas (N_2) flow rates 1.6 and 8 L/min, respectively). UV-vis spectra were recorded on a Thermo Scientific Waltham, MA, USA Evolution 260 Bio spectrophotometer. Melting points were determined with a Gallenkamp apparatus.

Diamidocalix[4]arene **1** was synthesized according to a literature procedure [30].

3.2. Synthesis of Compound **2**

A solution of calixarene **1** (0.3 g, 0.46 mmol) in a mixture of THF (25 mL) and pyridine (12 mL) was added dropwise to a solution of 4-nitroaniline (0.25 g, 1.81 mmol) and $NaNO_2$ (0.25 g, 3.62 mmol) in 2M HCl (20 mL) cooled to 0 °C with an ice bath. The mixture was stirred at 0 °C for 2 h, then 1M HCl (20 mL) was added, and the resulting precipitate was filtered on a Buchner funnel and washed with methanol. Recrystallization of the solid from CH_2Cl_2-methanol yielded compound **2** as a dark orange powder in 36% yield (0.157 g, 0.16 mmol). M.p. > 300 °C (dec.) ^{1}H NMR (400 MHz, CDCl$_3$): δ 8.34 (d, J = 8.9 Hz, 4H, ArH$_{o\text{-NO2}}$), 7.93 (d, J = 8.9 Hz, 4H, ArH $_{m\text{-NO2}}$), 7.75 (s, 4H, ArH), 7.12 (d, J = 7.6 Hz,

4H, ArH), 6.87 (t, J = 7.6 Hz, 2H, ArH), 4.92 (s, 4H, OCH_2), 4.67 (d, J = 13.1 Hz, 4H, $ArCH_2Ar$), 3.57–3.50 (m, 8H, $ArCH_2Ar$ and NCH_2), 3.42 (q, J = 7.2 Hz, 4H, NCH_2), 1.31 (t, J = 7.1 Hz, 6H, CH_3), 1.24 (t, J = 7.1 Hz, 6H, CH_3). ^{13}C NMR (100 MHz, $CDCl_3$): δ 167.4 (C=O), 158.2 (ArH), 156.5 (ArH), 153.9 (ArH), 147.8 (ArH), 145.6 (ArH), 133.4 (ArH), 129.5 (ArH), 128.9 (ArH), 125.7 (ArH), 124.8 (ArH), 124.7 (ArH), 122.8 (ArH), 73.1 (OCH_2), 41.0 (NCH_2), 40.3 (NCH_2), 31.7 ($ArCH_2Ar$), 14.3 (CH_3), 13.0 (CH_3). ESI-MS (m/z): calcd. for $C_{52}H_{52}N_8O_{10}Na^+$ ([M + Na]$^+$) 971.37, found 971.62; calcd. for $C_{52}H_{52}N_8O_{10}K^+$ ([M + K]$^+$) 987.34, found 987.61. ([M + K$^+$]).

3.3. X-ray Data Collection

Single crystal data were collected with a Bruker D8 PhotonII area detector diffractometer using microfocus radiation sources (Mo Kα: λ = 0.71073 Å, and Cu Kα: λ = 1.54178 Å). Complete datasets were obtained by merging several series of exposure frames collected at 200 K. An absorption correction was applied with the program SADABS [32] for **2**, **2·CaCl$_2$** and **2·Ca-A**. Data measured for **2·Ca-B** were indexed, integrated with a twinning matrix, and scaled using CrysAlis Pro software (version 42.49) accessed on: 12 May 2023 [33]. The structures were solved with ShelxT [34] and refined on F^2 with full-matrix least squares (ShelxL [35]), using the Olex2 software package (version 1.5) [36]. Non-hydrogen atoms were refined with anisotropic thermal parameters for all compounds.

In **2**, both amide and phenylazo moieties were found disordered over two distinct sites (65:35 occupancy) and refined with a series of SIMU and DFIX restraints. The hydrogen bonded to phenol oxygen was located from the residual electron density map and refined with DFIX restraint.

The asymmetric unit of **2·CaCl$_2$** comprised the calcium cation complexed by the calixarene, one chloride anion positioned on an inversion center, a second chloride anion disordered over two sites having site occupancy factors of 0.5 each, and a third chloride anion disordered over two sites and exchanging hydrogen bond interactions with disordered methanol molecules. According to the refinement, two chloride anions were present overall. One coordinated methanol molecule was split over two sites and refined with DFIX, DANG, and SIMU restraints. The hydrogen atoms bound to the nitrogen atoms of the azo moieties (N5 and N2) could be located from the difference Fourier map, and they were then placed at the calculated positions and refined. The other two partially occupied methanol molecules were pinpointed in the asymmetric unit, while the residual electron density (112 electrons in a volume of 504 Å^3 for the unit cell) was modelled according to the MASK program. Hence, the masked electron density was taken into account as eight methanol molecules of crystallization for the unit cell.

In **2·Ca-A**, one of the phenylazo-4-phenol moiety was found disordered over two distinct sites, refined with a 60:40 site occupancy factor and employing a series of SIMU restraints.

In **2·Ca-B**, owing to poor data quality and twinning, HKL5 was used for the final refinement, and several DFIX, DANG, SIMU, and ISOR restraints and AFIX66 constraints were required. The MASK program was used to model the residual electron density, which corresponded to 290 electrons and 1137 Å^3 per unit cell. The volume and residual electron count were consistent with the presence of four methanol molecules of crystallization per formula unit.

Crystal data and structure refinement details are provided in the supporting information, Table S1. CCDC 2261451–2261454 contains the supplementary crystallographic data for this paper. These data can be obtained free of charge from the Cambridge Crystallographic Data Centre via www.ccdc.cam.ac.uk/structures accessed on: 12 May 2023.

4. Conclusions

A new member of the azocalix[4]arenes family (**2**) was conveniently synthesized in only two steps from commercially available calix[4]arene and studied as a chromoionophore for the Ca^{2+} cation. Thanks to a combination of solution phase and solid-state investigation techniques, we have been able to thoroughly characterize the **2**·Ca^{2+} complex in light of the azo-phenol–quinone-hydrazone tautomeric equilibration both in neutral and basic conditions. In acetonitrile, the complexation of Ca^{2+} induces a shift of the tautomeric equilibration of the azo-phenol groups towards the quinone-hydrazone form without deprotonation of the ligand. The same structure is observed in the solid state, where, in addition, the counteranion Cl^{-} is hydrogen bonded to the hydrazone NH group. The addition of a weak base such as TEA to the acetonitrile solution of the complex results in the deprotonation of the two quinone-hydrazone moieties, which revert to the azo-phenolate tautomers. The X-ray structure of the neutral complex obtained by treating ligand **2** with $Ca(OH)_2$ confirms the restoration of the double N=N bond and the consequent localization of the negative charge on the phenolate ring, in close contact with the cation.

Due to the widespread use of the phenylazophenol moiety as a reporter unit in chromoionophores, we believe our results will be useful for the design of new optical sensors for metal cations.

Supplementary Materials: The following supporting information can be downloaded at: https://www.mdpi.com/article/10.3390/molecules28124704/s1p; Scheme S1: Resonance forms of azo-phenol moiety; Figure S1: ^{1}H NMR spectra of ligand **2** upon addition of $Ca(ClO_4)_2$ and TEA; Table S1: Crystal data and structure refinement details for the measured compounds; Figure S2: Thermal ellipsoid representation of **2**; Figure S3: Thermal ellipsoid representation of **2·CaCl₂**; Figure S4: Thermal ellipsoid representation of **2·Ca-A**; Figure S5: Thermal ellipsoid representation of **2·Ca-B**; Figure S6: Packing view of **2**; Figure S7: View of intermolecular interactions for **2**; Table S2: Selected intra and intermolecular interactions for **2**; Figure S8: View of **2·CaCl₂** showing hydrogen bond interactions involving chloride anions; Figure S9: View of **2·CaCl₂** showing supramolecular interactions; Table S3: Selected supramolecular interactions for **2·CaCl₂**; Figure S10: View of **2·Ca-A** showing supramolecular interactions; Table S4: Selected supramolecular interactions for **2·Ca-A**; Figure S11: Coordination bond lengths and geometries for Ca complexes **2·CaCl₂**, **2·Ca-A** and **2·Ca-B**; Figure S12: Selected bond lengths for **2** and **2·CaCl₂**; Figure S13: Selected bond lengths for **2·Ca-A**.

Author Contributions: Conceptualization, L.B.; methodology, L.B.; validation, L.B., A.C. and L.M.; formal analysis, D.B.; investigation, L.B. and D.B.; resources, A.C. and L.M.; data curation, D.B.; writing—original draft preparation, L.B.; writing—review and editing, L.B., A.C. and L.M.; visualization, D.B.; supervision, L.B.; project administration, L.B.; funding acquisition, A.C. and L.M. All authors have read and agreed to the published version of the manuscript.

Funding: This research received no external funding.

Institutional Review Board Statement: Not applicable.

Data Availability Statement: Data sets are available from the corresponding author on request.

Acknowledgments: This work benefited from the equipment and support of the COMP-HUB Initiative, the "Departments of Excellence" program of the Italian Ministry for Education, University and Research (MIUR, 2018–2022). The "Centro Interfacoltà di Misure" (CIM) of the University of Parma is acknowledged for the use of their NMR spectrometer. Chiesi Farmaceutici SpA is acknowledged for the support of the D8 Venture X-ray equipment.

Conflicts of Interest: The authors declare no conflict of interest.

Sample Availability: Samples of compound **2** are available from the corresponding author.

References

1. Lohr, H.G.; Vogtle, F. Chromo- and Fluoroionophores. A New Class of Dye Reagents. *Acc. Chem. Res.* **1985**, *18*, 65–72. [CrossRef]
2. Ludwig, R. Turning Ionophores into Chromo- and Fluoro-Ionophores. In *Calixarenes 2001*; Asfari, Z., Böhmer, V., Harrowfield, J., Vicens, J., Saadioui, M., Eds.; Springer: Dordrecht, The Netherlands, 2001; pp. 598–611.
3. Diamond, D.; McKervey, M.A. Calixarene-Based Sensing Agents. *Chem. Soc. Rev.* **1996**, *25*, 15–24. [CrossRef]
4. Na Kim, H.; Xiu Ren, W.; Seung Kim, J.; Yoon, J. Fluorescent and Colorimetric Sensors for Detection of Lead, Cadmium, and Mercury Ions. *Chem. Soc. Rev.* **2012**, *41*, 3210–3244. [CrossRef] [PubMed]
5. Choi, M.J.; Kim, M.Y.; Chang, S.K. A New Hg^{2+}-Selective Chromoionophore Based on Calix[4]arenediazacrown Ether. *Chem. Commun.* **2001**, *1*, 1664–1665. [CrossRef] [PubMed]

6. Kim, E.J.; Choe, J.I.; Chang, S.K. Novel Ca^{2+}-Selective Merocyanine-Type Chromoionophore Derived from Calix[4]arene-diamide. *Tetrahedron Lett.* **2003**, *44*, 5299–5302. [CrossRef]

7. Chang, K.C.; Su, I.H.; Lee, G.H.; Chung, W.S. Triazole- and Azo-Coupled Calix[4]arene as a Highly Sensitive Chromogenic Sensor for Ca^{2+} and Pb^{2+} Ions. *Tetrahedron Lett.* **2007**, *48*, 7274–7278. [CrossRef]

8. Song, K.C.; Choi, M.G.; Ryu, D.H.; Kim, K.N.; Chang, S.K. Ratiometric Chemosensing of Mg^{2+} Ions by a Calix[4]arene Diamide Derivative. *Tetrahedron Lett.* **2007**, *48*, 5397–5400. [CrossRef]

9. Chawla, H.M.; Sahu, S.N. Synthesis of Novel Chromogenic Azocalix[4]arenemonoquinones and Their Binding with Alkali Metal Cations. *J. Incl. Phenom. Macrocycl. Chem.* **2009**, *63*, 141–149. [CrossRef]

10. Shinkai, S.; Araki, K.; Shibata, J.; Tsugawa, D.; Manabe, O. Diazo-Coupling Reactions with Calix[4]arene. pKa Determination with Chromophoric Azocalix[4]Arenes. *Chem. Lett.* **1989**, *18*, 931–934. [CrossRef]

11. Oueslati, F.; Dumazet-Bonnamour, I.; Lamartine, R. New Chromogenic Azocalix[4]arene Podands Incorporating 2,2'-Bipyridyl Subunits. *New J. Chem.* **2003**, *27*, 644–647. [CrossRef]

12. Rouis, A.; Mlika, R.; Dridi, C.; Davenas, J.; Ben Ouada, H.; Halouani, H.; Bonnamour, I.; Jaffrezic, N. Optical Spectroscopy Studies of the Complexation of Chromogenic Azo-Calix[4]arene with Eu^{3+}, Ag+ and Cu^{2+} Ions. *Mater. Sci. Eng. C* **2006**, *26*, 247–252. [CrossRef]

13. Chen, Y.J.; Chung, W.S. Tetrazoles and Para-Substituted Phenylazo-Coupled Calix[4]arenes as Highly Sensitive Chromogenic Sensors for Ca^{2+}. *Eur. J. Org. Chem.* **2009**, *2009*, 4770–4776. [CrossRef]

14. Wang, J.; Guan, H.; Ge, C.; Fan, P.; Xing, X.; Shang, Y. Azocalix[4]arene with Three Distal Ethyl Ester Residues as a Highly Selective Chromogenic Sensor for Ca^{2+} Ions. *Heterocycl. Commun.* **2018**, *24*, 147–150. [CrossRef]

15. Shimizu, H.; Iwamoto, K.; Fujimoto, K.; Shinkai, S. Chromogenic Calix[4]arene. *Chem. Lett.* **1991**, *20*, 2147–2150. [CrossRef]

16. Gordon, J.L.M.; Böhmer, V.; Vogt, W. A Calixarene-Based Chromoionophore for the Larger Alkali Metals. *Tetrahedron Lett.* **1995**, *36*, 2445–2448. [CrossRef]

17. Hayashita, T.; Kunogi, K.; Yamamoto, H.; Shinkai, S. Selective Colorimetry of Sodium Ion in Acidic Aqueous Media by Calix[4]crown Chromoionophore. *Anal. Sci.* **1997**, *13*, 161–166. [CrossRef]

18. Kim, N.Y.; Chang, S.-K. Calix[4]arenes Bearing Two Distal Azophenol Moieties: Highly Selective Chromogenic Ionophores for the Recognition of Ca^{2+} Ion. *J. Org. Chem.* **1998**, *63*, 2362–2364. [CrossRef]

19. Van Der Veen, N.J.; Egberink, R.J.M.; Engbersen, J.F.J.; Van Veggel, F.J.C.M.; Reinhoudt, D.N. Conformationally Flexible Calix[4]arene Chromoionophores: Optical Transduction of Soft Metal Ion Complexation by Cation-π Interactions. *Chem. Commun.* **1999**, 681–682. [CrossRef]

20. Halouani, H.; Dumazet-Bonnamour, I.; Duchamp, C.; Bavoux, C.; Ehlinger, N.; Perrin, M.; Lamartine, R. Synthesis, Conformations and Extraction Properties of New Chromogenic Calix[4]arene Amide Derivatives. *Eur. J. Org. Chem.* **2002**, 4202–4210. [CrossRef]

21. Kim, J.Y.; Kim, G.; Kim, C.R.; Lee, S.H.; Lee, J.H.; Kim, J.S. UV Band Splitting of Chromogenic Azo-Coupled Calix[4]crown upon Cation Complexation. *J. Org. Chem.* **2003**, *68*, 1933–1937. [CrossRef]

22. Kishimoto, S.; Kitahara, S.; Manabe, O.; Hiyama, H. Tautomerism and Dissociation of 4-Arylazo-1-Naphthols in Various Solvents. *J. Org. Chem.* **1978**, *43*, 3882–3886. [CrossRef]

23. Stoyanov, S.; Antonov, L.; Soloveytchik, B.; Petrova, V. Quantitative Analysis of Tautomeric Equilibrium in 1-Phenylazo-4-Naphthols-a New Approach. *Dye. Pigment.* **1994**, *26*, 149–158. [CrossRef]

24. Chapoteau, E.; Czech, B.P.; Gebauer, C.R.; Kumar, A.; Leong, K.; Mytych, D.T.; Zazulak, W.; Desai, D.H.; Luboch, E.; Krzykawski, J.; et al. Phenylazophenol-Quinone Phenylhydrazone Tautomerism in Chromogenic Cryptands and Corands with Inward-Facing Phenolic Units and Their Acyclic Analogues. *J. Org. Chem.* **1991**, *56*, 2575–2579. [CrossRef]

25. Joshi, H.; Kamounah, F.S.; van der Zwan, G.; Gooijer, C.; Antonov, L. Temperature Dependent Absorption Spectroscopy of Some Tautomeric Azo Dyes and Schiff Bases. *J. Chem. Soc. Perkin Trans. 2* **2001**, *12*, 2303–2308. [CrossRef]

26. Kim, T.H.; Kim, S.H.; Van Tan, L.; Dong, Y.; Kim, H.; Kim, J.S. Diazo-Coupled Calix[4]arenes for Qualitative Analytical Screening of Metal Ions. *Talanta* **2008**, *74*, 1654–1658. [CrossRef] [PubMed]

27. Kim, T.H. Spectrophotometric and Electrochemical Study for Metal Ion Binding of Azocalix[4]arene Bearing p-Ethylester Group. *Spectrochim. Acta Part A Mol. Biomol. Spectrosc.* **2017**, *178*, 8–13. [CrossRef]

28. Beer, P.D.; Drew, M.G.B.; Leeson, P.B.; Ogden, M.I. Metal Complexes of a Calix[4]arene Diamide: Syntheses, Crystal Structures and Molecular Mechanics Calculations on [Fe(L1-2H)][FeCl4] and [Er(L1-2H)(Picrate)] (L1 = 5,11,17,23-Tetra-Tert-Butyl-25,27-Bis(Diethylcarbamoylmethoxy)Calix[4]Arene). *Inorg. Chim. Acta* **1996**, *246*, 133–141. [CrossRef]

29. Beer, P.D.; Drew, M.G.B.; Kan, M.; Leeson, P.B.; Ogden, M.I.; Williams, G. Lanthanide Structures, Coordination, and Extraction Investigations of a 1,3-Bis(Diethyl Amide)-Substituted Calix[4]arene Ligand. *Inorg. Chem.* **1996**, *35*, 2202–2211. [CrossRef]

30. Casnati, A.; Fischer, C.; Guardigli, M.; Isernia, A.; Manet, I.; Sabbatini, N.; Ungaro, R. Synthesis of Calix[4]arene Receptors Incorporating (2,2'-Bipyridin-6-Yl)Methyl and (9-Methyl-1,10-Phenanthrolin-2-Yl)Methyl Chromophores and Luminescence of Their Eu^{3+} and Tb^{3+} Complexes. *J. Chem. Soc. Perkin Trans. 2* **1996**, *3*, 395–399. [CrossRef]

31. Chawla, H.M.; Sahu, S.N.; Shrivastava, R. Synthesis and Binding Characteristics of Novel Calix[4]arene(amidocrown) Diquinones. *Can. J. Chem.* **2009**, *87*, 523–531. [CrossRef]

32. Krause, L.; Herbst-Irmer, R.; Sheldrick, G.M.; Stalke, D. Comparison of Silver and Molybdenum Microfocus X-Ray Sources for Single-Crystal Structure Determination. *J. Appl. Crystallogr.* **2015**, *48*, 3–10. [CrossRef] [PubMed]

33. *CrysAlis Pro Software*; Version 42.49; Agilent Technologies Ltd.: Oxfordshire, UK, 2013.

34. Sheldrick, G.M. SHELXT—Integrated Space-Group and Crystal-Structure Determination. *Acta Crystallogr. Sect. A Found. Crystallogr.* **2015**, *71*, 3–8. [CrossRef] [PubMed]
35. Sheldrick, G.M. Crystal Structure Refinement with SHELXL. *Acta Crystallogr. Sect. C Struct. Chem.* **2015**, *71*, 3–8. [CrossRef] [PubMed]
36. Dolomanov, O.V.; Bourhis, L.J.; Gildea, R.J.; Howard, J.A.K.; Puschmann, H. OLEX2: A Complete Structure Solution, Refinement and Analysis Program. *J. Appl. Crystallogr.* **2009**, *42*, 339–341. [CrossRef]

Article

Hexahexyloxycalix[6]arene, a Conformationally Adaptive Host for the Complexation of Linear and Branched Alkylammonium Guests

Veronica Iuliano, Carmen Talotta *, Paolo Della Sala, Margherita De Rosa, Annunziata Soriente, Placido Neri and Carmine Gaeta *

Dipartimento di Chimica e Biologia "A. Zambelli", Università di Salerno, Via Giovanni Paolo II 132, I-84084 Salerno, Italy; viuliano@unisa.it (V.I.); pdellasala@unisa.it (P.D.S.); maderosa@unisa.it (M.D.R.); titti@unisa.it (A.S.); neri@unisa.it (P.N.)
* Correspondence: ctalotta@unisa.it (C.T.); cgaeta@unisa.it (C.G.)

Abstract: Hexahexyloxycalix[6]arene **2b** leads to the *endo*-cavity complexation of linear and branched alkylammonium guests showing a conformational adaptive behavior in CDCl$_3$ solution. Linear *n*-pentylammonium guest **6a⁺** induces the cone conformation of **2b** at the expense of the 1,2,3-alternate, which is the most abundant conformer of **2b** in the absence of a guest. In a different way, branched alkylammonium guests, such as *tert*-butylammonium **6b⁺** and isopropylammonium **6c⁺**, select the 1,2,3-alternate as the favored **2b** conformation (**6b⁺**/**6c⁺**⊂**2b**$^{1,2,3-alt}$), but other complexes in which **2b** adopts different conformations, namely, **6b⁺**/**6c⁺**⊂**2b**^{cone}, **6b⁺**/**6c⁺**⊂**2b**^{paco}, and **6b⁺**/**6c⁺**⊂**2b**$^{1,2-alt}$, have also been revealed. Binding constant values determined via NMR experiments indicated that the 1,2,3-alternate was the best-fitting **2b** conformation for the complexation of branched alkylammonium guests, followed by cone > paco > 1,2-alt. Our NCI and NBO calculations suggest that the H-bonding interactions (⁺N–H⋯O) between the ammonium group of the guest and the oxygen atoms of calixarene **2b** are the main determinants of the stability order of the four complexes. These interactions are weakened by increasing the guest steric encumbrance, thus leading to a lower binding affinity. Two stabilizing H-bonds are possible with the 1,2,3-alt- and cone-**2b** conformations, whereas only one H-bond is possible with the other paco- and 1,2-alt-**2b** stereoisomers.

Keywords: conformation; calixarene; molecular recognition; alkylammonium guests

Citation: Iuliano, V.; Talotta, C.; Della Sala, P.; De Rosa, M.; Soriente, A.; Neri, P.; Gaeta, C. Hexahexyloxycalix[6]arene, a Conformationally Adaptive Host for the Complexation of Linear and Branched Alkylammonium Guests. *Molecules* **2023**, *28*, 4749. https://doi.org/10.3390/molecules28124749

Academic Editor: Gheorghe Dan Pantoş

Received: 9 May 2023
Revised: 5 June 2023
Accepted: 8 June 2023
Published: 13 June 2023

1. Introduction

Molecular recognition [1] is a fundamental process in living systems, and due to our understanding of the secondary interactions that stabilize the ligand@protein complexes, it has been possible to design novel biomimetic guest@host supramolecular systems. A natural process such as protein–substrate binding often involves conformational changes, which can occur prior to the binding event ('*conformational selection model*' [2,3]) or during the binding event ('*induced-fit model*' [4,5]). According to the *conformational selection model*, the protein conformational changes may take place prior to ligand binding, and then the stabilization of a specific protein structure is caused by its complexation with the substrate. In contrast, in the *induced-fit binding model*, the conformational change takes place upon substrate binding [6].

The design of artificial ammonium receptors is an exciting topic of research in supramolecular chemistry, which takes inspiration from natural systems [7,8]. Thus, the recognition of ammonium guests by macrocycles such as calixarenes [9], pillararenes [10,11], prismarenes [12–14], cucurbiturils [15], naphthotubes [16], oxatubarene [17], cycloparaphenylenes [18], and saucerarenes [19], has received substantial attention in recent years. In particular, very recently, Jiang [17] reported a new class of macrocycles named oxatub[4]arenes, which showed biomimetic conformational adaptation behavior [20,21]. In fact, oxatubarenes can take on four interconvertible conformations through the flipping of naphthalene rings. Jiang showed that according to the "conformational selection model", a specific ammonium guest can select the best-fitting oxatubarene conformer, altering the initial equilibrium distribution of the conformers. Calix[2]naphth[2]arene [22] macrocycle, which we reported in 2020, is composed of two phenol and two naphthalene rings and can adopt five potential conformations, but the 1,2-alternate conformation is the only one that achieves the best binding when alkali metal cations are present.

Additionally, calix[5]arene macrocycles can show conformational response to ammonium guests [23]. In fact, very recently, we showed that the cone and partial cone are the best-fitting conformers of a calix[5]arene for secondary ammonium cations.

In 2010 [24], our group showed that conformationally mobile hexamethoxycalix[6]arene macrocycle **2a** forms pseudorotaxane complexes via threading with dialkylammonium cations coupled with the weakly coordinating barfate anion (BArF⁻ = tetrakis [3,5-bis (trifluoromethyl)phenyl]borate) (Chart 1). The *endo*-cavity complexation of linear and branched alkylammonium cations, as barfate salts, was also observed with calix[5]arene **1** [23], doubly bridged calix[7]arene **3** [9], dihomooxacalix[4]arene **4** [25,26], and calix[4]arene **5** [27].

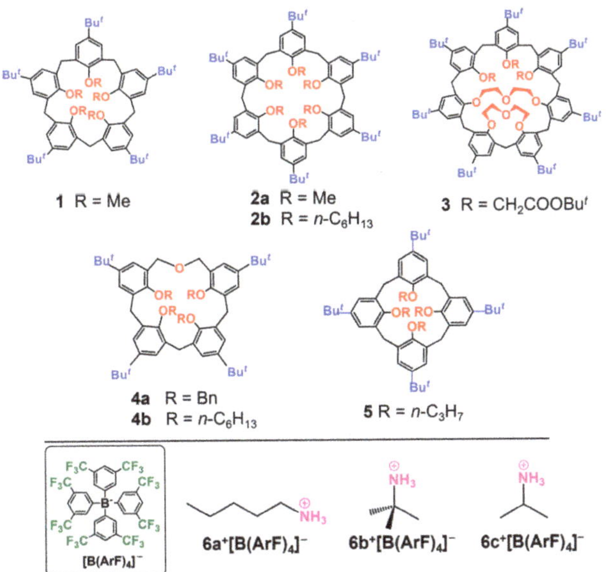

Chart 1. Structures of calix[*n*]arene derivatives **1–5** and alkylammonium **6a−d⁺**[B(ArF)₄]⁻ salts.

The calix[6]arene macrocycle can adopt eight distinct conformations (Figure 1), including cone, partial cone, 1,2-alternate, 1,3-alternate, 1,4-alternate, 1,2,3-alternate, 1,2,4-alternate, and 1,3,5-alternate [28–30], and conformational interconversion between them occurs by means of *"oxygen-through-the-annulus"* or/and *"tert-butyl-through-the-annulus"* passage.

As previously reported by us [30], the ^{1}H NMR spectrum of hexahexyloxycalix[6]arene **2b** in CDCl₃ at 298 K (Figure 2a) shows broad signals indicative of a slow conformational mobility of the macrocycle with respect to the NMR time scale. We showed that after lowering the temperature to 233 K, the ^{1}H NMR resonances of **2b** decoalesced to form sharp signals compatible with the presence of 1,2,3-alternate (favored) and cone conformations of **2b**. Based on these considerations, hexahexyloxycalix[6]arene **2b** is the ideal candidate for studying the conformational response of the calix[6]arene skeleton to the presence of ammonium guests. The question to address now is specifically whether linear and branched alkylammonium cations **6a–c⁺** can form complexes with **2b,** as well as the possibility of complexation-induced selection of the 1,2,3-alternate/cone or alternative conformations of hexahexyloxycalix[6]arene **2b**.

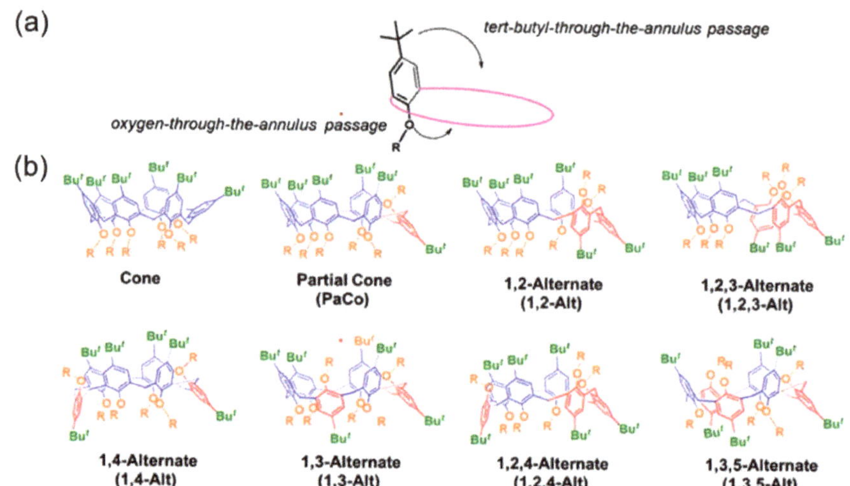

Figure 1. (**a**) Possible pathways for conformational interconversion in *p-tert*-butylcalix[6]arenes. (**b**) The eight basic conformations of calix[6]arene derivatives.

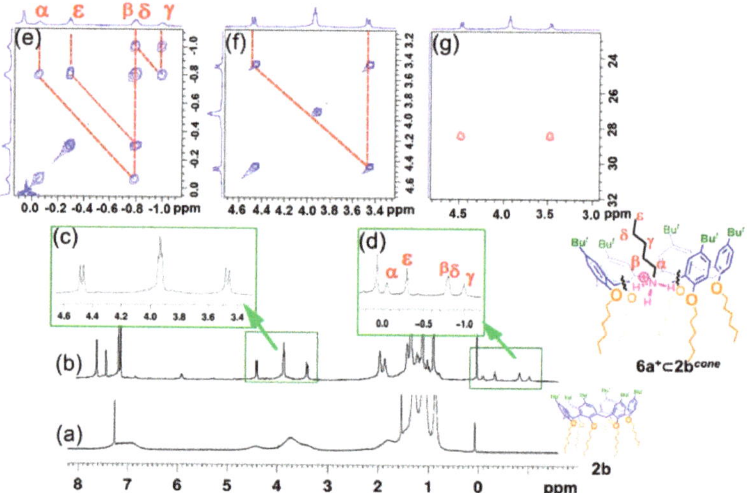

Figure 2. 1D and 2D NMR spectra (CDCl₃, 600 MHz, 298 K) for the complexation of **6a⁺** by **2b**. (**a**) ¹H NMR spectrum of **2b**; (**b**) ¹H NMR spectrum of an equimolar solution of **2b** and **6a⁺**[B(ArF)₄]⁻ (3 mM); (**c,d**) enlargements of two different portions of (**b**); (**e,f**) portions of the COSY-45; and (**g**) HSQC spectrum (CDCl₃, 600 MHz, 298 K) of the solution in (**b**).

Consequently, prompted by these considerations, we decided to investigate the molecular recognition properties of **2b** toward linear and branched alkylammonium ions **6a–c⁺** as barfate salts [B(ArF)₄]⁻ (Chart 1).

2. Results and Discussion

2.1. Binding Ability of **2b** toward n-Pentylammonium Guest **6a⁺**[B(ArF)₄]⁻

The complexation ability of **2b** toward **6a⁺**[B(ArF)₄]⁻ (Chart 1) was investigated at 298 K via 1D and 2D NMR experiments. The ¹H NMR spectrum of an equimolar (3 mM) solution of **2b** and **6a⁺**[B(ArF)₄]⁻ in CDCl₃ at 298 K showed typical features [24] of the presence of the *endo*-cavity **6a**$^+$⊂**2b** complex (Scheme 1). In particular, the formation of the complex was ascertained based on the appearance of a new set of slowly exchanging signals in the up-field negative region of the spectrum (Figure 2) attributable to the *n*-pentyl chain of **6a**$^+$ shielded inside the cavity of the calix[6]arene macrocycle.

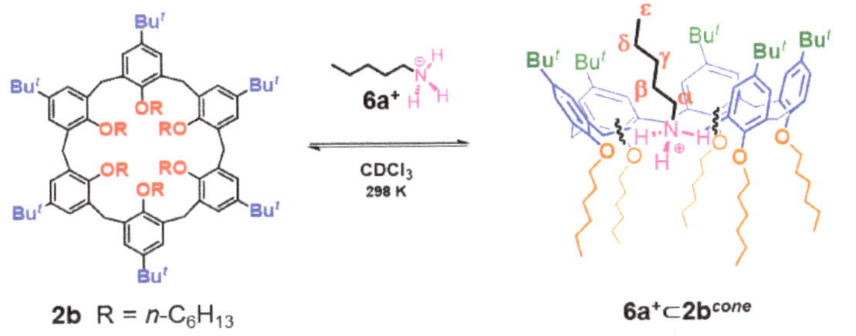

Scheme 1. Formation of the **6a**$^+$⊂**2b**^{cone} complex.

The NMR signals of **6a**$^+$ complexed inside the cavity of **2b** were assigned via a COSY-45 experiment (Figure 2e). As a result, the NH$_3$$^+$ signal at 5.99 ppm correlated with the α-protons at –0.06 ppm, which coupled with the β-methylene group at –0.81 ppm. This, in turn, showed a cross peak with the γ-protons at –0.99 ppm. The γ-protons were coupled with the δ-methylene group at –0.81 ppm, which was correlated with the ε-methyl group at –0.28 ppm. The COSY-45 spectrum of the **6a**$^+$⊂**2b** complex revealed the presence of an AX system (Figure 2f) at 3.47/4.48 ppm (Δδ = 0.99 ppm), which correlated in the HSQC spectrum with a ^{13}C resonance at 28.4 ppm, attributable to the ArCH$_2$Ar groups. In agreement with Gutsche's "^{1}H NMR" rule [28,30] and the "^{13}C NMR single rule" of de Mendoza [29,30], these results are only compatible with the formation of the *endo*-cavity **6a**$^+$⊂**2b**^{cone} complex, in which the calix[6]arene adopts a cone conformation. In conclusion, the pentylammonium cation **6a**$^+$ selected the cone as the best-fitting **2b** conformation at the expense of the 1,2,3-alternate, which was the most abundant species in the initial conformational equilibria of **2b** [30].

The calculation of the binding constant for the formation of the **6a**$^+$⊂**2b**^{cone} complex through direct peak integration was not possible, as the ^{1}H NMR signals (Figure 2b) of the free **2b** and guest **6a**$^+$ were not detected in the ^{1}H NMR spectrum of their 1:1 mixture, shown in Figure 2b. Consequently, a binding constant calculation was performed by means of a competition experiment in which 1 equiv of pentylammonium **6a**$^+$ was mixed with a 1:1 mixture of **2b** and *n*-butylammonium (in CDCl$_3$) as barfate salt. Previously, we reported on the formation of the *n*-**BuNH$_3$**$^+$⊂**2b**^{cone} complex in CDCl$_3$ with a K_{ass} value of 8.3 ± 0.1 × 10^6 M^{-1} [12,13,24,31]. After the mixing of **2b**, *n*-**BuNH$_3$**$^+$, and **6a**$^+$ (1/1/1 molar ratio), the *n*-**BuNH$_3$**$^+$⊂**2b**^{cone} complex was preferentially formed over the **6a**$^+$⊂**2b**^{cone} one in a 1.6:1.0 ratio. Thus, from these data, a binding constant of 5.2 ± 0.1 × 10^6 M^{-1} was calculated for the formation of the **6a**$^+$⊂**2b**^{cone} complex in CDCl$_3$ at 298 K (see the Supplementary Materials for further details).

The DFT-optimized structure of the **6a**$^+$⊂**2b**^{cone} complex (Figure 3), calculated on the B3LYP/6-31G(d,p) theoretical level, revealed that the N$^+$ atom of **6a**$^+$ sits 0.30 Å above the mean plane of the ethereal oxygen atoms of **2b**. H-bonding interactions were detected between the $^+$NH$_3$—ammonium group of **6a**$^+$ and the oxygen atoms of **2b**, with a $^+$N···O^{2b} average distance of 2.83 Å and an average $^+$N-H··· O^{2b} angle of 165.1°. In addition, CH···π interactions were detected between the aromatic rings of **2b** and the pentyl chain of **6a**$^+$ confined inside the cavity of **2b** (Figure 2a–c), with an average C–H···π^{centroid} distance of 2.99 Å. A second-order perturbation theory SOPT analysis [32] of the Fock matrix in the NBO [33] basis and NCI (non-covalent interactions) studies were conducted in order to rationalize the energy contribution of secondary interactions.

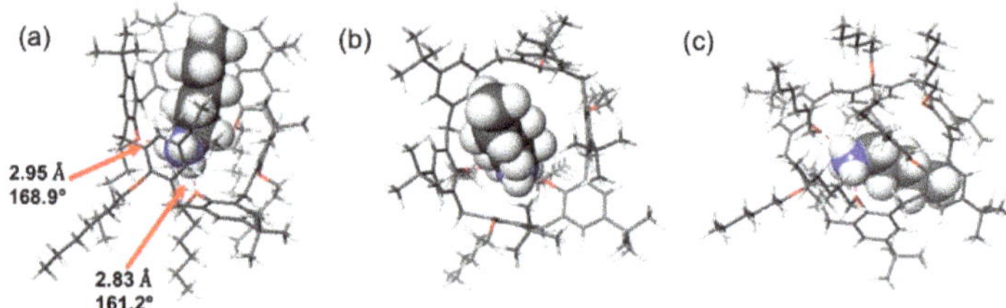

Figure 3. (a–c) Different views of the DFT-optimized structure of the **6a**⁺ ⊂ **2b**^*cone* complex. (a) Marked in red, ⁺N–H···O, H-bonding mean distance and angle.

Interestingly, the SOPT analysis conducted on **6a**⁺ ⊂ **2b**^*cone* indicated that a network of ⁺N–H···O^*2b* hydrogen bonding and C-H···π interactions stabilized the complex. We identified two lone-pair (LP) interactions between the oxygen atoms of **2b** and the N–H antibonding orbitals (i.e., H-bonding interactions, Figure 3) of **6a**⁺ in the **6a**⁺ ⊂ **2b** complex, which provided a 78% energetic contribution to the total stabilization energy of the complex.

2.2. Binding Ability of **2b** toward Tert-Butylammonium Guest **6b**⁺[B(ArF)₄]⁻

With these results in hand, we focused our attention on a branched alkylammonium guest such as *tert*-butylammonium **6b**⁺[B(ArF)₄]⁻, which revealed a very different behaviour compared to **6a**⁺[B(ArF)₄]⁻. Close inspection of the ¹H NMR spectrum (CDCl₃, 600 MHz, 298 K) of the equimolar mixture **2b**/**6b**⁺ (3 mM) in Figure 4 showed typical signals indicative of the *endo*-cavity complexation of **6b**⁺ inside the cavity of **2b** (Scheme 2).

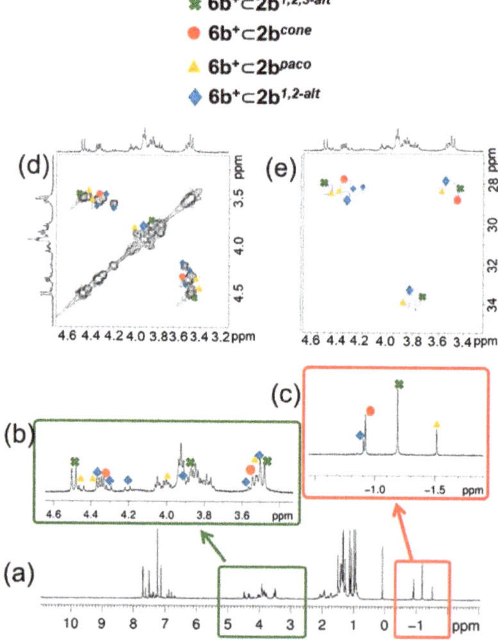

Figure 4. The 1D and 2D NMR spectra (CDCl₃, 600 MHz, 298 K) for the complexation of **6b**⁺ by **2b**. (a) ¹H NMR spectrum of an equimolar solution of **2b** and **6b**⁺[B(ArF)₄]⁻ (3 mM). (b,c) Enlargements of two different portions of (a). (d,e) Portions of the COSY-45 and HSQC spectra (CDCl₃, 600 MHz, 298 K) of the solution in (a).

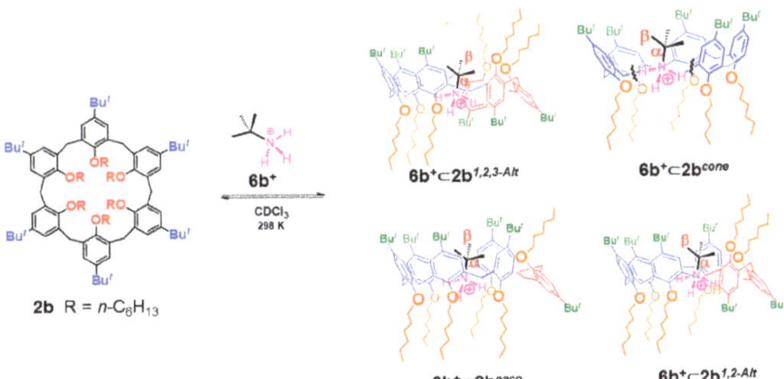

Scheme 2. *Endo*-cavity complexation of *tert*-butylammonium **6b⁺** as barfate salt [B(ArF)₄]⁻ by **2b**: formation of **6b⁺⊂2b**^{1,2,3-Alt}, **6b⁺⊂2b**^{cone}, **6b⁺⊂2b**^{paco}, and **6b⁺⊂2b**^{1,2-Alt} complexes.

Closer inspection of the methylene region (4.5–3.4 ppm) in the COSY-45 spectrum of the equimolar mixture **2b**/**6b⁺** (3 mM) showed the presence of several AX systems (marked in red, blue, yellow, and green in Figure 4), which correspond to the formation of various **6b⁺⊂2b** complexes in which the calix[6]arene macrocycle adopts different conformations. The 1D and 2D NMR analyses of the equimolar mixture **2b**/**6b⁺** (3 mM), supported with DFT calculations, made it possible to identify four complexes, namely, **6b⁺⊂2b**^{1,2,3-alt}, **6b⁺⊂2b**^{cone}, **6b⁺⊂2b**^{paco}, and **6b⁺⊂2b**^{1,2-alt} (Scheme 2), in which the calix[6]arene host adopts the 1,2,3-alternate, cone, partial cone, and 1,2-alternate conformations, respectively.

The integration of the ¹H NMR signals attributable to **6b⁺⊂2b**^{1,2,3-alt}, **6b⁺⊂2b**^{cone}, **6b⁺⊂2b**^{paco}, and **6b⁺⊂2b**^{1,2-alt} revealed that the four complexes were present in a 6:3:2:1 ratio. From these data and through a quantitative ¹H NMR experiment, using trichloroethylene (TCE) as the internal standard (see the experimental section for further details), the following apparent association constants were calculated: $2.6 \pm 0.2 \times 10^3$ M⁻¹ (**6b⁺⊂2b**^{1,2,3-alt}); $1.8 \pm 0.2 \times 10^3$ M⁻¹ (**6b⁺⊂2b**^{cone}), $8.1 \pm 0.2 \times 10^2$ M⁻¹ (**6b⁺⊂2b**^{paco}), and $4.3 \pm 0.2 \times 10^2$ M⁻¹ (**6b⁺⊂2b**^{1,2-alt}).

These values clearly demonstrated that the binding of the branched *tert*-butylammonium cation is generally less favored than that of linear *n*-pentylammonium, probably due to the greater steric encumbrance of the *t*-Bu group.

For the sake of clarity, hereafter, we report on the analysis of the 1D and 2D NMR spectra in Figure 4, supporting the identification of the four complexes formed upon mixing **2b** and **6b⁺** in an equimolar ratio (3 mM) in CDCl₃ at 298 K.

6b⁺⊂2b^{1,2,3-alt}: The presence of an ArCH₂Ar AX system at 3.49/4.49 ppm and an AB system at 3.86/3.77 ppm provides evidence of the **6b⁺⊂2b**^{1,2,3-alt} complex, in which the calix[6]arene adopts a 1,2,3-alternate conformation. The HSQC correlations of these ArCH₂Ar signals with carbon resonances at 28.3 and 33.7 ppm (Figure 4e) confirmed the presence of the calixarene host in the 1,2,3-alternate conformation, according to Gutsche's "¹H NMR" rule [28,30] and the "¹³C NMR single rule" of de Mendoza [29,30]. Finally, the *tert*-butyl singlet of **6b⁺** shielded inside the cavity of **2b**^{1,2,3-alt} was detected at −1.15 ppm (marked in green), a value significantly up-field-shifted as compared to the analogous signals of **6b⁺** hosted inside the cone-shaped cavity of **6b⁺⊂2b**^{cone} (−0.92 ppm, marked in red) and the **6b⁺⊂2b**^{1,2-alt} complex (−0.91 ppm, marked in blue) (*vide infra*).

6b⁺⊂2b^{cone}: The methylene region of the ¹H NMR spectrum in Figure 4a,b showed the presence of an AX system (COSY spectrum, indicated in red in Figure 4d) at 3.52/4.34 ppm ($\Delta\delta$ = 0.82 ppm), which correlated in the HSQC spectrum (Figure 4e) with a carbon resonance at 28.0 ppm, attributable to carbon atoms between *syn*-oriented aromatic rings. These signals can be assigned to the **6b⁺⊂2b**^{cone} complex, in which calix[6]arene **2b** adopts the cone conformation. Interestingly, the *tert*-butyl signal of **6b⁺** shielded inside the aromatic cavity of **2b**^{cone} was found at −0.92 ppm (marked in red in Figure 4c).

6b$^+$⊂2b^{paco}: The ^{1}H NMR spectrum in Figure 4a,b showed the presence of two less intense AX systems (COSY spectrum, Figure 4d) in a 1:1 ratio at 3.48/4.45 ppm ($\Delta\delta$ = 0.97 ppm) and 3.51/4.35 ppm ($\Delta\delta$ = 0.84 ppm), which correlated in the HSQC spectrum (Figure 4e) with carbon resonances at 28.3 and 28.6 ppm, respectively, attributable to ArCH$_2$Ar carbon atoms between *syn*-oriented aryl rings. In addition, an AB system (COSY spectrum) was detected at 3.86/3.99 ppm, attributable to ArCH$_2$Ar groups between *anti*-oriented aryl rings, which correlated with a carbon resonance at 33.8 ppm. According to the application of Gutsche's and de Mendoza's rules, these data were indicative of the presence of a **6b$^+$⊂2b^{paco}** complex in which the calix[6]arene adopted the partial cone conformation. Additionally, in this case, a singlet was detected at −1.51 ppm (marked in yellow in Figure 4), attributable to the -C(CH$_3$)$_3$ group of 6b+ shielded inside the cavity of 2b^{paco}.

6b$^+$⊂2b$^{1,2\text{-}alt}$: The presence of the **6b$^+$⊂2b$^{1,2\text{-}alt}$** complex was confirmed by three ArCH$_2$Ar AX systems (COSY spectrum, Figure 4d) at 3.49/4.30, 3.51/4.20, and 3.51/4.36 ppm ($\Delta\delta$ = 0.81, 0.69, and 0.85 ppm, respectively) in a 1:1:2 ratio, which correlated in the HSQC spectrum with carbon signals at 28.5, 28.6, and 28.4 ppm, respectively, and an AB ArCH$_2$Ar system at 3.86/3.96 ppm, which correlated with a carbon resonance at 33.6 ppm. In this case, in the negative region of the ^{1}H NMR, we observed the *tert*-butyl singlet of **6b$^+$** at −0.90 ppm (marked in blue in Figure 4c).

To investigate the energy contribution of noncovalent interactions, a SOPT analysis of the Fock matrix in the NBO basis was carried out on the DFT-optimized structures of the four **6b$^+$⊂2b** complexes.

The DFT-optimized structure of the **6b$^+$⊂2b$^{1,2,3\text{-}alt}$** complex indicated the presence of stabilizing H-bonding and C−H⋯π interactions between the guest **6b$^+$** and **2b$^{1,2,3\text{-}alt}$**. A SOPT analysis indicated that the stabilization energy was mainly due to the formation of two $^+$N–H⋯O^{2b} H-bonding interactions, which contributed 80% of the total binding energy (Table 1). This value is significantly higher than that calculated for the $^+$N−H⋯O^{2b} H−bonding interactions of the **6b$^+$⊂2b^{cone}** complex, which was 61% of the total energy of non-covalent interactions (Table 1) for this complex. The DFT-optimized structure of the **6b$^+$⊂2b^{cone}** complex calculated on the B3LYP/6-31G(d,p) theoretical level (Figure 5a) showed the presence of two $^+$N−H⋯O interactions with an average $^+$N⋯O^{2b} distance of 2.95 Å, being longer and weaker than that calculated for the **6b$^+$⊂2b$^{1,2,3\text{-}alt}$** complex ($^+$N⋯O^{2b} average distance of 2.80 Å), while an average $^+$N−H⋯O^{2b} angle of 165.1° was calculated for the **6b$^+$⊂2b^{cone}** complex (163.3° calculated for **6b$^+$⊂2b$^{1,2,3\text{-}alt}$**). Additionally, C−H⋯π interactions were detected between the *tert*-butyl group of 6b+ and the aromatic rings of **2b**, with an average C-H⋯π centroid distance of 2.95 Å and an average C-H⋯π centroid angle of 153.3°.

Table 1. Contribution of $^+$N–H⋯O hydrogen-bonding interactions to the total binding energy, as calculated via SOPT analysis of the Fock matrix in the natural bond orbital (NBO) basis for the complexes between **6b$^+$** and **2b** in different conformations.

Complex	K_{ass}	$^+$N–H⋯O H-Bonding Percentage of Total Binding Energy (%)
6b$^+$⊂2b$^{1,2,3\text{-}alt}$	$2.6 \pm 0.2 \times 10^3$ M^{-1}	80
6b$^+$⊂2b^{cone}	$1.8 \pm 0.2 \times 10^3$ M^{-1}	61
6b$^+$⊂2b^{paco}	$8.1 \pm 0.2 \times 10^2$ M^{-1}	53
6b$^+$⊂2b$^{1,2\text{-}alt}$	$4.3 \pm 0.2 \times 10^2$ M^{-1}	36

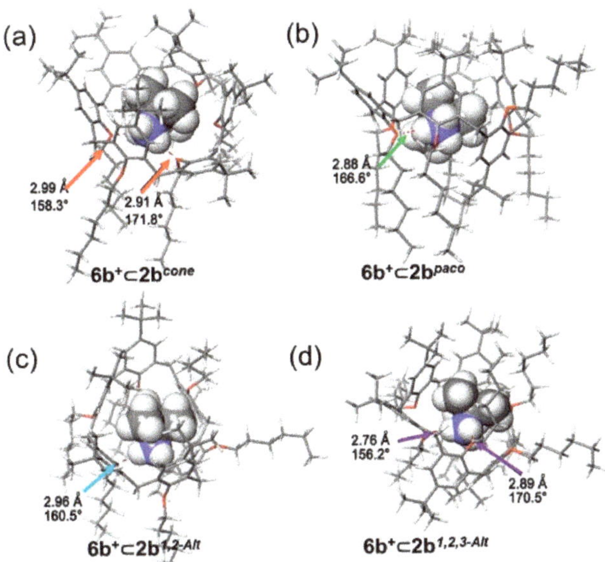

Figure 5. DFT-optimized structures of the complexes **6b⁺⊂2b**$^{1,2,3\text{-alt}}$ (a), **6b⁺⊂2b**^{cone} (b), **6b⁺⊂2b**^{paco} (c), and **6b⁺⊂2b**$^{1,2\text{-alt}}$ (d) calculated on the B3LYP/6-31G(d,p) theoretical level.

Concerning the DFT-optimized structure of **6b⁺⊂2b**^{paco} (Figure 5b), a single H-bonding interaction was detected between the NH_3^+ group of 6b+ and an oxygen atom of **2b**^{paco}, with a distance of 2.88 Å and an angle of 166.6°. The SOPT analysis revealed a lower value for the contribution of the $^+$N–H$\cdots$O^{2b} hydrogen-bonding interaction (53%, Table 1).

Similarly, the DFT-optimized structure of the **6b⁺⊂2b**$^{1,2\text{-alt}}$ complex (Figure 5c) suggested the existence of a single H-bonding interaction between the ammonium group of 6b+ and an oxygen atom of **2b**$^{1,2\text{-alt}}$. Here, the SOPT analysis also revealed a lower 36% contribution of the $^+$N–H$\cdots$O^{2b} hydrogen-bonding interaction (Table 1) to the total binding energy.

These results clearly indicate that the *tert*-butylammonium guest **6b⁺** prefers the 1,2,3-alt–**2b** as the best-fitting host conformation to a greater extent than the other conformations. In addition, the NCI analysis suggests that the stabilization induced by the H-bonding interactions between the ammonium group of **6b⁺** and the oxygen atoms of the calixarene **2b** ($^+$N–H$\cdots$O) plays a crucial role in determining the thermodynamic stabilities of the four complexes shown in Scheme 2. A careful comparison of the DFT-optimized structures of the **6a⁺⊂2b** and **6b⁺⊂2b** complexes reveals significant differences in the binding modes of the two guests. In particular, the greater steric requirements of the *tert*-butyl group of **6b⁺** force it to occupy a deeper position inside the calix[6]arene cavity (in each of the four conformations), leading to greater deformation of the host, which, in turn, implies a higher energetical cost and a lower binding constant. In this way, the deeply positioned **6b⁺** guest is able to form two stabilizing H-bonds with the 1,2,3-alt- and cone-**2b** conformations, whereas only one H-bond is possible with the other paco- and 1,2-alt-**2b** isomers.

2.3. Binding Ability of **2b** toward Isopropylammonium Guest **6c⁺**[B(ArF)₄]⁻

The *endo*-cavity complexation of the isopropylammonium guest **6c⁺**[B(ArF)₄]⁻ showed very similar features to those observed in the complexation of the *tert*-butylammonium guest **6b⁺** with **2b**. In fact, the 1D and 2D NMR analysis of an equimolar mixture of **6c⁺**[B(ArF)₄]⁻ and **2b** resulted in the formation of a well-defined mixture of stereoisomeric complexes in which the calix[6]arene adopted different conformations, namely,

6c⁺⊂2b$^{1,2,3\text{-}alt}$, **6c⁺⊂2b**^{cone}, **6c⁺⊂2b**^{paco}, and **6c⁺⊂2b**$^{1,2\text{-}alt}$ (Figure 6). The four complexes were found in a 9:3:2:1 ratio according to the integration of their ^{1}H NMR signals.

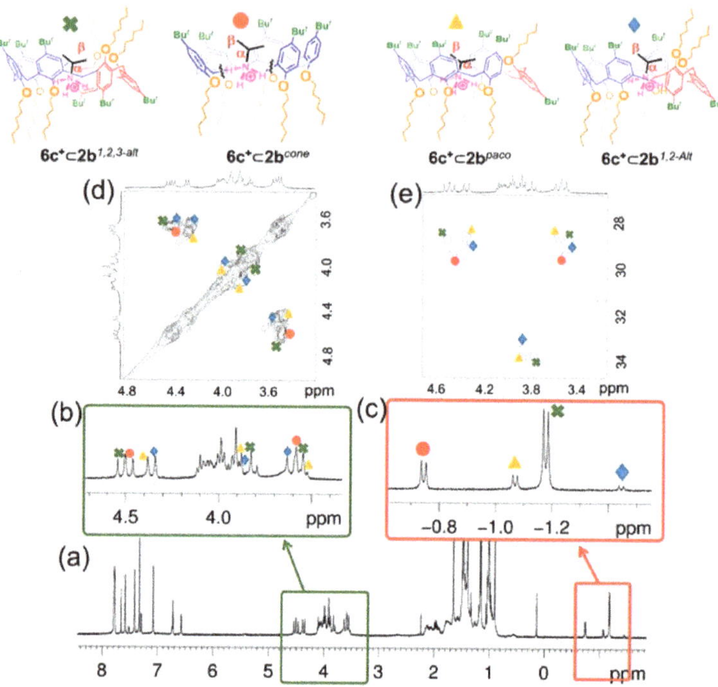

Figure 6. The 1D and 2D NMR spectra (CDCl$_3$, 600 MHz, 298 K): (**a**) ^{1}H NMR spectrum of an equimolar solution of **2b** and **6b⁺**[B(ArF)$_4$]$^-$ (3 mM); (**b**,**c**) enlargements of two different portions of (**a**); (**d**,**e**) portions of the COSY-45 and HSQC spectra (CDCl$_3$, 600 MHz, 298 K) of the solution in (**a**).

Through a quantitative ^{1}H NMR (CDCl$_3$, 600 MHz, 298 K) experiment, using TCE as the internal standard, four binding constants were calculated for the slowly exchanging complexes: $3.1 \pm 0.2 \times 10^4$ M^{-1} for **6c⁺⊂2b**$^{1,2,3\text{-}alt}$, $9.1 \pm 0.2 \times 10^3$ M^{-1} for **6c⁺⊂2b**^{cone}, $6.1 \pm 0.2 \times 10^3$ M^{-1} for **6c⁺⊂2b**^{paco}, and $1.2 \pm 0.2 \times 10^3$ M^{-1} for **6c⁺⊂2b**$^{1,2\text{-}alt}$ (SI).

These results clearly confirmed the role of the steric encumbrance, since the binding of the more branched *tert*-butylammonium cation is less favored than that of the less branched *i*-propylammonium, which, in turn, is less favored with respect to linear *n*-pentylammonium.

The ^{1}H NMR spectrum (600 MHz, 298 K, Figure 6a–c) of a 1:1 mixture of **6c⁺**[B(ArF)$_4$]$^-$and **2b** (3 mM) in CDCl$_3$ showed a series of doublets shielded at the negative values of chemical shifts and attributable to *endo*-cavity complexed β–CH$_3$ groups of 6c+: -1.50 (*J* = 6.8 Hz), -1.23 (*J* = 6.8 Hz), -1.13 (*J* = 6.8 Hz), -0.68 (*J* = 6.8 Hz) ppm, assigned to **6c⁺⊂2b**$^{1,2\text{-}alt}$, **6c⁺⊂2b**$^{1,2,3\text{-}alt}$, **6c⁺⊂2b**^{paco}, and **6c⁺⊂2b**^{cone}, respectively.

6c⁺⊂2b$^{1,2,3\text{-}alt}$: The formation of the **6c⁺⊂2b**$^{1,2,3\text{-}alt}$ complex, in which the calix[6]arene adopts a 1,2,3-alternate conformation, was ascertained according to the presence of an AX system at 3.47/4.46 ppm ($\Delta\delta$ = 1.00 ppm; HSQC: ^1J correlation with an Ar^{13}CH$_2$Ar signal at 28.3 ppm) and an AB system at 3.73/3.81 ppm ($\Delta\delta$ = 0.08 ppm; HSQC: ^1J correlation with an ArCH$_2$Ar signal at 34.0 ppm, marked in green in Figure 6).

6b⁺⊂2b^{cone}: The formation of the **6b⁺⊂2b**^{cone} complex was established based on the presence of an AX system at 3.48/4.41 ppm, which correlated in the HSQC spectrum (Figure 6e) with a carbon resonance at 28.0 ppm, attributable to carbon atoms between *syn*-oriented aromatic rings (red marked in Figure 6).

$6c^+ \subset 2b^{paco}$: The ^{1}H NMR and COSY-45 spectra of the $6c^+ \subset 2b^{paco}$ complex showed two AX systems at 3.54/4.29 ppm ($\Delta\delta = 0.75$) and 3.53/4.29 ppm ($\Delta\delta = 0.76$) in 1:1 ratio, which correlated in the HSQC spectrum (Figure 6e) with carbon resonances at 28.3 and 28.6 ppm, respectively. Moreover, the HSQC spectrum of the complex showed a correlation between a carbon resonance at 34.0 ppm and an AB system at 3.81/4.01) ppm, attributable to $ArCH_2Ar$ groups between *anti*-oriented Ar rings.

$6c^+ \subset 2b^{paco}$: The presence of three $ArCH_2Ar$ AX systems (COSY spectrum, Figure 6d) at 3.47/4.40, 3.52/4.34, and 3.57/4.20 ppm ($\Delta\delta = 0.93$, 0.82, and 0.63 ppm, respectively) in a 1:1:2 ratio (indicated in blue in Figure 6) is compatible with the presence of the $6c^+ \subset 2b^{1,2-alt}$ complex.

The H-bonding contribution to the total binding energy calculated for the four $6c^+ \subset 2b$ complexes (Table 2) is in agreement with their thermodynamic stability order evaluated based on K_{ass} values in $CDCl_3$ at 298 K: $6c^+ \subset 2b^{1,2,3-alt} > 6c^+ \subset 2b^{cone} > 6c^+ \subset 2b^{paco} > 6c^+ \subset 2b^{1,2-alt}$. These results clearly indicate that, once again, the 1,2,3-alt-2b was selected as the best-fitting host conformation in the presence of a branched alkylammonium guest such as $6c^+$. The stability order of the $6c^+ \subset 2b$ complexes is in full agreement with that observed in the presence of the *tert*-butylammonium guest 6b+. The natural bond orbital and noncovalent interaction analyses indicate that H-bonding interactions between the ammonium group of $6c^+$ and the oxygen atoms of the calixarene 2b ($^+$N–H$\cdots$O) provide the key stabilization factor for the $6c^+ \subset 2b$ complexes. In fact, they account for 87%, 77%, 66%, and 52% of the total binding energy for the $6c^+ \subset 2b^{1,2,3-alt}$, $6c^+ \subset 2b^{cone}$, $6c^+ \subset 2b^{paco}$, and $6c^+ \subset 2b^{1,2-alt}$ complexes, respectively (Table 2).

Table 2. Contribution of $^+$N–H$\cdots$O hydrogen-bonding interactions to the total binding energy, as calculated via SOPT analysis of the Fock matrix in the natural bond orbital (NBO) basis for the complexes between 6c+ and 2b in different conformations.

Complex	K_{ass}	$^+$N–H$\cdots$O H-Bonding Percentage of Total Binding Energy (%)
$6c^+ \subset 2b^{1,2,3-alt}$	$3.1 \pm 0.2 \times 10^4$ M^{-1}	87
$6c^+ \subset 2b^{cone}$	$9.1 \pm 0.2 \times 10^3$ M^{-1}	77
$6c^+ \subset 2b^{paco}$	$6.1 \pm 0.2 \times 10^3$ M^{-1}	66
$6c^+ \subset 2b^{1,2-alt}$	$1.2 \pm 0.2 \times 10^3$ M^{-1}	52

At this point, it was important to verify whether the thermodynamic stability of the individual conformational complexes was in accordance with the Expanding Coefficient (EC) parameter recently proposed by our group [34]. In fact, it was found that the EC parameter can be conveniently correlated with the thermodynamic stability of supramolecular complexes obeying the induced-fit or conformational selection models and governed by weak secondary interactions. EC is defined as the ratio between the volume of the host cavity after complexation and that of the host cavity before complexation [34].

The EC values were thus calculated from the cavity volumes of the complexed and free host using the above DFT-optimized structures and that of the 1,2,3-alternate ground 2b host for all the $6b^+ \subset 2b$ and $6c^+ \subset 2b$ complexes (SI) with the Caver software [35–37]. In detail, taking the $6c^+ \subset 2b$ complexes as an example, the EC values of 6.03, 6.37, 4.84, and 6.03 were found for $6c^+ \subset 2b^{1,2,3-alt}$, $6c^+ \subset 2b^{cone}$, $6c^+ \subset 2b^{paco}$, and $6c^+ \subset 2b^{1,2-alt}$, respectively, whose log K_{app} values were 4.49, 3.96, 3.79, and 3.08, respectively (SI). As is clearly evident, the previously observed linear correlation between the EC and log K_{app} [34] is not a factor here [30]. This can be easily explained by considering the number of H-bonding interactions, which, as stated above, is the main determinant of the thermodynamic stability of these complexes. Thus, if we only consider those complexes with two H-bonding interactions, $6c^+ \subset 2b^{1,2,3-alt}$ (EC = 6.03, log K_{app} = 4.49) and $6c^+ \subset 2b^{cone}$, (EC = 6.37, log K_{app} = 3.96), we find a good correlation between an increasing EC and decreasing log K_{app}. In the same way, considering only those complexes with one H-bonding interaction, $6c^+ \subset 2b^{paco}$ (EC = 4.84, log K_{app} = 3.79) and $6c^+ \subset 2b^{1,2\text{-}alt}$ (EC = 6.03, log K_{app} = 3.08), we again find a good correlation between an increasing EC and decreasing log K_{app}. Therefore, these findings are in accordance with our previous statement [34] that "the EC parameter can be considered of general applicability in all those instances in which no new strong intermolecular interactions (e.g., H-bonds) are generated during the induced-fit process".

3. Conclusions

This study clearly shows that hexahexyloxycalix[6]arene **2b** can complex alkylammonium guests and shows a conformational adaptive behavior. Thus, in the presence of *n*-pentylammonium $6a^+$, the cone-**2b** is the best-fitting conformation at the expense of the 1,2,3-alternate-**2b**, which is the most abundant conformer in the absence of a guest. In a different way, branched alkylammonium guests, such as *tert*-butylammonium $6b^+$ and isopropylammonium $6c^+$, select a combination of conformers of **2b**. Four complexes were revealed and characterized using 1D and 2D NMR spectra, in which **2b** adopted different conformations, namely, $6b^+/6c^+ \subset 2b^{1,2,3\text{-}alt}$, $6b^+/6c^+ \subset 2b^{cone}$, $6b^+/6c^+ \subset 2b^{paco}$, and $6b^+/6c^+ \subset 2b^{1,2\text{-}alt}$. The binding constant values determined through NMR experiments indicated that the 1,2,3-alternate was the best-fitting **2b** conformation for the complexation of branched alkylammonium guests, followed by cone > paco > 1,2-alt. The NCI and NBO calculations suggest that the H-bonding interactions between the ammonium group of the guest and the oxygen atoms of the calixarene **2b** ($^+$N–H$\cdots$O) played a crucial role in determining the thermodynamic stability order of the four complexes. In addition, it was found that an increase in branching from *n*-Pent to *i*-Pr and in *t*-Bu groups leads to a corresponding decrease in binding affinity. Therefore, it can be concluded that higher steric encumbrance leads to weaker H-bonding interactions and, hence, to a lower binding energy.

Supplementary Materials: The following supporting information can be downloaded at: https://www.mdpi.com/article/10.3390/molecules28124749/s1, Supplementary Materials containing the procedure for the formation of alkylammonium@calixarene complexes; 1D and 2D NMR spectra of complexes (Figures S1–S21); details of DFT, NBO/NCI, calculations and stability constant determination [38].

Author Contributions: Conceptualization, C.G., C.T. and P.N.; methodology and investigation, V.I.; data curation, P.D.S.; writing—original draft preparation, V.I. and P.D.S.; writing—review and editing, P.N., C.T. and C.G.; visualization, A.S. and M.D.R.; supervision, P.N. and C.G.; project administration, P.N. All authors have read and agreed to the published version of the manuscript.

Funding: This research was funded by the UNIVERSITY of SALERNO, FARB 2021.

Data Availability Statement: Not applicable.

Conflicts of Interest: The authors declare no conflict of interest.

References

1. Atwood, J.L.; Davies, J.E.D.; Macnicol, D.D.; Vogtle, F. *Comprehensive Supramolecular Chemistry*; Pergamon Press: New York, NY, USA, 1996.
2. Hammes, G.G.; Chang, Y.-C.; Oas, T.G. Conformational selection or induced fit: A flux description of reaction mechanism. *Proc. Natl. Acad. Sci. USA* **2009**, *106*, 13737–13741. [CrossRef] [PubMed]
3. Monod, J.; Wyman, J.; Changeux, J.-P. On the nature of allosteric transitions: A plausible model. *J. Mol. Biol.* **1965**, *12*, 88–188. [CrossRef] [PubMed]
4. Paul, F.; Weikl, T.R. How to Distinguish Conformational Selection and Induced Fit Based on Chemical Relaxation Rates. *PLoS Comput. Biol.* **2016**, *12*, e1005067. [CrossRef] [PubMed]
5. Koshland, D.E. Application of a Theory of Enzyme Specificity to Protein Synthesis. *Proc. Natl. Acad. Sci. USA* **1958**, *44*, 98–104. [CrossRef]
6. Takahashi, D.T.; Gadelle, D.; Agama, K.; Kiselev, E.; Zhang, H.; Yab, E.; Petrella, S.; Forterre, P.; Pommier, Y.; Mayer, C. Topoisomerase I (TOP1) Dynamics: Conformational Transition from Open to Closed States. *Nat. Commun.* **2022**, *13*, 59. [CrossRef]
7. Miles, T.F.; Bower, K.S.; Lester, H.A.; Dougherty, D.A. A Coupled Array of Noncovalent Interactions Impacts the Function of the 5-HT$_3$A Serotonin Receptor in an Agonist-Specific Way. *ACS Chem. Neurosci.* **2012**, *3*, 753–760. [CrossRef]
8. Duffy, N.H.; Lester, H.A.; Dougherty, D.A. Ondansetron and Granisetron Binding Orientation in the 5-HT$_3$ Receptor Determined by Unnatural Amino Acid Mutagenesis. *ACS Chem. Biol.* **2012**, *7*, 1738–1745. [CrossRef]

9.   Gaeta, C.; Talotta, C.; Farina, F.; Campi, G.; Camalli, M.; Neri, P. Conformational Features and Recognition Properties of a Confo mationally Blocked Calix[7]Arene Derivative. *Chem. Eur. J.* **2012**, *18*, 1219–1230. [CrossRef] [PubMed]
10.   Fan, J.; Chen, Y.; Cao, D.; Yang, Y.-W.; Jia, X.; Li, C. Host–Guest Properties of Pillar[7]Arene towards Substituted Adamantane Ammonium Cations. *RSC Adv.* **2014**, *4*, 4330–4333. [CrossRef]
11.   Li, C.; Shu, X.; Li, J.; Fan, J.; Chen, Z.; Weng, L.; Jia, X. Selective and Effective Binding of Pillar[5,6]Arenes toward Secondary Ammonium Salts with a Weakly Coordinating Counteranion. *Org. Lett.* **2012**, *14*, 4126–4129. [CrossRef]
12.   Della Sala, P.; Del Regno, R.; Talotta, C.; Capobianco, A.; Hickey, N.; Geremia, S.; De Rosa, M.; Spinella, A.; Soriente, A.; Neri, P.; et al. Prismarenes: A New Class of Macrocyclic Hosts Obtained by Templation in a Thermodynamically Controlled Synthesis. *J. Am. Chem. Soc.* **2020**, *142*, 1752–1756. [CrossRef]
13.   Della Sala, P.; Del Regno, R.; Di Marino, L.; Calabrese, C.; Palo, C.; Talotta, C.; Geremia, S.; Hickey, N.; Capobianco, A.; Neri, P.; et al. An Intramolecularly Self-Templated Synthesis of Macrocycles: Self-Filling Effects on the Formation of Prismarenes. *Chem. Sci.* **2021**, *12*, 9952–9961. [CrossRef] [PubMed]
14.   Della Sala, P.; Del Regno, R.; Iuliano, V.; Capobianco, A.; Talotta, C.; Geremia, S.; Hickey, N.; Neri, P.; Gaeta, C. Confused Prism[5]arene: A Conformationally Adaptive Host by Stereoselective Opening of the 1,4-Bridged Naphthalene Flap. *Chem. Eur. J.* **2023**, *29*, e202203030. [CrossRef]
15.   Mock, W.L.; Shih, N.Y. Structure and Selectivity in Host-Guest Complexes of Cucurbituril. *J. Org. Chem.* **1986**, *51*, 4440–4446. [CrossRef]
16.   Yang, L.-P.; Wang, X.; Yao, H.; Jiang, W. Naphthotubes: Macrocyclic Hosts with a Biomimetic Cavity Feature. *Acc. Chem. Res.* **2020**, *53*, 198–208. [CrossRef] [PubMed]
17.   Jia, F.; He, Z.; Yang, L.-P.; Pan, Z.-S.; Yi, M.; Jiang, R.-W.; Jiang, W. Oxatub[4]Arene: A Smart Macrocyclic Receptor with Multiple Interconvertible Cavities. *Chem. Sci.* **2015**, *6*, 6731–6738. [CrossRef]
18.   Della Sala, P.; Talotta, C.; Caruso, T.; De Rosa, M.; Soriente, A.; Neri, P.; Gaeta, C. Tuning Cycloparaphenylene Host Properties by Chemical Modification. *J. Org. Chem.* **2017**, *82*, 9885–9889. [CrossRef]
19.   Li, J.; Zhou, H.; Han, Y.; Chen, C. Saucer[*n*]arenes: Synthesis, Structure, Complexation, and Guest-Induced Circularly Polarized Luminescence Property. *Angew. Chem. Int. Ed.* **2021**, *60*, 21927–21933. [CrossRef]
20.   Jia, F.; Li, D.; He, S.; Yang, L.; Jiang, W. Conformational Effects on the Threading Kinetics of Dumbbell-Shaped Guests into the Cavity of Oxatub[4]Arene. *Angew. Chem. Int. Ed.* **2022**, *61*, e2022123. [CrossRef]
21.   Wang, X.; Jia, F.; Yang, L.-P.; Zhou, H.; Jiang, W. Conformationally Adaptive Macrocycles with Flipping Aromatic Sidewalls. *Chem. Soc. Rev.* **2020**, *49*, 4176–4188. [CrossRef]
22.   Del Regno, R.; Della Sala, P.; Spinella, A.; Talotta, C.; Iannone, D.; Geremia, S.; Hickey, N.; Neri, P.; Gaeta, C. Calix[2]Naphth[2]Arene: A Class of Naphthalene–Phenol Hybrid Macrocyclic Hosts. *Org. Lett.* **2020**, *22*, 6166–6170. [CrossRef] [PubMed]
23.   De Rosa, M.; Talotta, C.; Gaeta, C.; Soriente, A.; Neri, P.; Pappalardo, S.; Gattuso, G.; Notti, A.; Parisi, M.F.; Pisagatti, I. Calix[5]arene through-the-Annulus Threading of Dialkylammonium Guests Weakly Paired to the TFPB Anion. *J. Org. Chem.* **2017**, *82*, 5162–5168. [CrossRef] [PubMed]
24.   Gaeta, C.; Troisi, F.; Neri, P. *Endo*-Cavity Complexation and Through-the-Annulus Threading of Large Calixarenes Induced by Very Loose Alkylammonium Ion Pairs. *Org. Lett.* **2010**, *12*, 2092–2095. [CrossRef]
25.   Gaeta, C.; Talotta, C.; Farina, F.; Teixeira, F.A.; Marcos, P.M.; Ascenso, J.R.; Neri, P. Alkylammonium Cation Complexation into the Narrow Cavity of Dihomooxacalix[4]Arene Macrocycle. *J. Org. Chem.* **2012**, *77*, 10285–10293. [CrossRef]
26.   Talotta, C.; Gaeta, C.; De Rosa, M.; Ascenso, J.R.; Marcos, P.M.; Neri, P. Alkylammonium Guest Induced-Fit Recognition by a Flexible Dihomo oxacalix[4]Arene Derivative: Alkylammonium Guest Induced-Fit Recognition. *Eur. J. Org. Chem.* **2016**, *2016*, 158–167. [CrossRef]
27.   Talotta, C.; Gaeta, C.; Neri, P. *Endo*-Complexation of Alkylammonium Ions by Calix[4]Arene Cavity: Facilitating Cation−π Inte actions through the Weakly Coordinating Anion Approach. *J. Org. Chem.* **2014**, *79*, 9842–9846. [CrossRef]
28.   Kanamathareddy, S.; Gutsche, C.D. Calixarenes. 29. Aroylation and Arylmethylation of Calix[6]Arenes. *J. Org. Chem.* **1992**, *57*, 3160–3166. [CrossRef]
29.   Jaime, C.; De Mendoza, J.; Prados, P.; Nieto, P.M.; Sanchez, C. Carbon-13 NMR Chemical Shifts. A Single Rule to Determine the Conformation of Calix[4]Arenes. *J. Org. Chem.* **1991**, *56*, 3372–3376. [CrossRef]
30.   Gaeta, C.; Talotta, C.; Neri, P. Calix[6]Arene-Based Atropoisomeric Pseudo[2]Rotaxanes. *Beilstein J. Org. Chem.* **2018**, *14*, 2112–2124. [CrossRef]
31.   Hirose, K. *Analytical Methods in Supramolecular Chemistry*; Schalley, C.A., Ed.; Wiley-VCH: Weinheim, Germany, 2007; Chapter 2; pp. 17–54.
32.   Weinhold, F.; Landis, C.R. *Valency and Bonding: A Natural Bond Orbital Donor-Acceptor Perspective*, 1st ed.; Cambridge University Press: Cambridge, UK, 2005.
33.   Johnson, E.R.; Keinan, S.; Mori-Sánchez, P.; Contreras-García, J.; Cohen, A.J.; Yang, W. Revealing Noncovalent Interactions. *J. Am. Chem. Soc.* **2010**, *132*, 6498–6506. [CrossRef]
34.   Talotta, C.; Concilio, G.; de Rosa, M.; Soriente, A.; Gaeta, C.; Rescifina, A.; Ballester, P.; Neri, P. Expanding coefficient: A parameter to assess the stability of induced-fit complexes. *Org. Lett.* **2021**, *23*, 1804–1808. [CrossRef] [PubMed]

35. Jurcik, A.; Bednar, D.; Byska, J.; Marques, S.M.; Furmanova, K.; Daniel, L.; Kokkonen, P.; Brezovsky, J.; Strnad, O.; Stourac, J.; et al. CAVER analyst 2.0: Analysis and visualization of channels and tunnels in protein structures and molecular dynamics trajectories. *Bioinformatics* **2018**, *34*, 3586–3588. [CrossRef] [PubMed]
36. Frisch, M.J.; Trucks, G.W.; Schlegel, H.B.; Scuseria, G.E.; Robb, M.A.; Cheeseman, J.R.; Scalmani, G.; Barone, V.; Mennucci, B.; Petersson, G.A.; et al. *Gaussian 16, Revision A.03*; Gaussian Inc.: Wallingford, CT, USA, 2016.
37. Krieger, E.; Vriend, G. YASARA View-molecular graphics for all devices-from smartphones to workstations. *Bioinformatics* **2014**, *30*, 2981–2982. [CrossRef] [PubMed]
38. Lu, T.; Chen, F.W. Multiwfn: A multifunctional wavefunction analyzer. *J. Comput. Chem.* **2012**, *33*, 580. [CrossRef]

Article

Detection of Aromatic Hydrocarbons in Aqueous Solutions Using Quartz Tuning Fork Sensors Modified with Calix[4]arene Methoxy Ester Self-Assembled Monolayers: Experimental and Density Functional Theory Study

Shofiur Rahman [1,*], Mahmoud A. Al-Gawati [1,2], Fatimah S. Alfaifi [2], Wadha Khalaf Alenazi [2], Nahed Alarifi [2], Hamad Albrithen [1,2], Abdullah N. Alodhayb [1,2,*] and Paris E. Georghiou [3,*]

1 Biological and Environmental Sensing Research Unit, King Abdullah Institute for Nanotechnology, King Saud University, P.O. Box 2455, Riyadh 11451, Saudi Arabia
2 Department of Physics and Astronomy, College of Science, King Saud University, P.O. Box 2455, Riyadh 11451, Saudi Arabia; falfaifi@ksu.edu.sa (F.S.A.)
3 Department of Chemistry, Memorial University of Newfoundland, St. John's, NL A1C 5S7, Canada
* Correspondence: mrahman1@ksu.edu.sa (S.R.); aalodhayb@ksu.edu.sa (A.N.A.); parisg@mun.ca (P.E.G.)

check for **updates**

Citation: Rahman, S.; Al-Gawati, M.A.; Alfaifi, F.S.; Alenazi, W.K.; Alarifi, N.; Albrithen, H.; Alodhayb, A.N.; Georghiou, P.E. Detection of Aromatic Hydrocarbons in Aqueous Solutions Using Quartz Tuning Fork Sensors Modified with Calix[4]arene Methoxy Ester Self-Assembled Monolayers: Experimental and Density Functional Theory Study. *Molecules* 2023, *28*, 6808. https://doi.org/10.3390/molecules28196808

Academic Editor: Paula M. Marcos

Received: 31 August 2023
Revised: 21 September 2023
Accepted: 22 September 2023
Published: 26 September 2023

Abstract: Quartz tuning forks (QTFs), which were coated with gold and with self-assembled monolayers (SAM) of a lower-rim functionalized calix[4]arene methoxy ester (CME), were used for the detection of benzene, toluene, and ethylbenzene in water samples. The QTF device was tested by measuring the respective frequency shifts obtained using small (100 μL) samples of aqueous benzene, toluene, and ethylbenzene at four different concentrations (10^{-12}, 10^{-10}, 10^{-8}, and 10^{-6} M). The QTFs had lower limits of detection for all three aromatic hydrocarbons in the 10^{-14} M range, with the highest resonance frequency shifts ($\pm5\%$) being shown for the corresponding 10^{-6} M solutions in the following order: benzene (199 Hz) > toluene (191 Hz) > ethylbenzene (149 Hz). The frequency shifts measured with the QTFs relative to that in deionized water were inversely proportional to the concentration/mass of the analytes. Insights into the effects of the alkyl groups of the aromatic hydrocarbons on the electronic interaction energies for their hypothetical 1:1 supramolecular host–guest binding with the CME sensing layer were obtained through density functional theory (DFT) calculations of the electronic interaction energies (ΔIEs) using B3LYP-D3/GenECP with a mixed basis set: LANL2DZ and 6-311++g(d,p), CAM-B3LYP/LANL2DZ, and PBE/LANL2DZ. The magnitudes of the ΔIEs were in the following order: [Au4-CME⊃[benzene] > [Au4-CME]⊃[toluene] > [Au4-CME]⊃[ethylbenzene]. The gas-phase BSSE-uncorrected ΔIE values for these complexes were higher, with values of -96.86, -87.80, and -79.33 kJ mol^{-1}, respectively, and -86.39, -77.23, and -67.63 kJ mol^{-1}, respectively, for the corresponding BSSE-corrected values using B3LYP-D3/GenECP with LANL2dZ and 6-311++g(d,p). The computational findings strongly support the experimental results, revealing the same trend in the ΔIEs for the proposed hypothetical binding modes between the tested analytes with the CME SAMs on the Au-QTF sensing surfaces.

Keywords: quartz tuning fork; calix[4]arene; self-assembled monolayer; aromatic hydrocarbons; density functional theory

1. Introduction

The aromatic hydrocarbons benzene, toluene, ethylbenzene, and xylene, which are collectively and commonly referred to as BTEX, are components of gasoline hydrocarbons. They are also part of the general grouping of volatile organic compounds (VOCs), which are found ubiquitously in our global environment [1]. BTEX compounds are released into the environment by a variety of different sources, which Yu et al. have classified as either being pyrogenic or petrogenic or arising from processed products [2]. As examples, BTEX compounds are released into the atmosphere as a result of the incomplete combustion of gasoline by motor vehicles and also from the combustion of organic materials or biomass, as well as other fossil fuels, including coal and crude oil. All of these are examples of pyrogenic sources. Oil seepages, which can occur naturally or from accidental oil spills, also discharge numerous VOCs, including BTEX [1,2], into the environment. These and other less-volatile polycyclic aromatic hydrocarbons (PAHs) can pose a significant threat both to the environment and directly to human health [3] in many parts of the world. These effects have been well researched and documented in the scientific literature [1–6]. Furthermore, crude oil processing has been reported to contribute up to 16% of hydrocarbon VOC emissions [7]. The preceding are all examples of petrogenic sources. In addition to being released into the atmosphere, many of these compounds appear in soil and potable water supplies, through the contamination of either aquifers or the water supply, e.g., lakes, rivers, or dams themselves. Although they are generally considered to be water-immiscible, the reported data on the water solubilities of benzene, toluene, and ethylbenzene are 1.80 g/L [8], 0.519 g/L [9], and 0.015 g/100 mL [10], respectively. These levels are significantly higher than the U.S. Environmental Protection Agency (EPA)'s maximum permissible levels of 5 μg/L, 1 mg/L, and 700 μg/L, respectively [11]. As an example of petrogenic environmental water contamination, Adeniran et al. provided a comprehensive review of groundwater contamination by crude oil spillages [12]. On the other hand, in addition to the fact that BTEX compounds enter the environment through the sources referred to above, other sources include industrial activities, and BTEXs have been found to be common groundwater contaminants at 1 to 3 μg/L [13]. This is because they are the most-frequently produced petrochemical intermediate chemicals, which, in addition to their use in gasoline, are also widely used as solvents in rubber, chemical, coating, dyeing, glue, printing, pesticides, paints, fuel additives, pharmaceutical industries, etc. These are examples of the third and "processed" sources of BTEX referred to by Yu et al [2]. Benzene is essential for producing products such as polystyrene from its ethylbenzene derivative, detergents, medications, paints, and pesticides, but it is particularly toxic by itself [14]. Ethylbenzene's dehydrogenation into styrene contributes to 85% of the total of styrene manufactured [15]. Toluene is extensively used in paint products [2], but it is not as toxic as benzene and ethylbenzene, so it is used indoors, where, initially, relatively high air levels can be measured [16].

Unfortunately, the concentrations of BTEX in the environment have increased significantly over the years, due to the main sources described above. This pollution has led to aquifer, groundwater, and surface-water contamination [2], which have also had a significant impact on marine organisms, even at low levels. Therefore, developing efficient, high-performance, and environmentally friendly techniques that allow the ultrasensitive and precise detection of BTEX on small samples without requiring significant preparation times is highly desirable.

Many analytical techniques have been used to quantify the concentration of benzene, toluene, and ethylbenzene in water at trace levels, including gas chromatography–mass spectrometry (GC-MS) [17], headspace-gas chromatography–mass spectrometry (HSGC-MS) [18], high-performance liquid chromatography (HPLC) [19], UV–laser-induced fluorescence spectroscopy [20], and polymer-based quartz crystal microbalance (QCM) sensors [21]. Although many of these methods have excellent detection limits and are capable of multi-element analysis, they also have some limitations in relation to the precise detection of trace amounts of aromatic compounds, such as benzene, toluene, and ethylbenzene. These limitations include expensive equipment facilities, knowledgeable and skilled personnel, and time-consuming and complex sample preparation.

Microcantilever sensors [22] and quartz tuning fork (QTF) sensors [23] based on micro-electromechanical systems (MEMSs) have been successfully used for detecting trace amounts of chemical species in aqueous media. These sensors can be used for various purposes, such as environmental monitoring, chemical warfare and the detection of explosives, and also, in medical research. It is important to note that each method has its own advantages and disadvantages, including differences in sensitivity, selectivity, cost, complexity, and the required sample preparation. The choice of method depends on the specific needs of the analysis, such as having a suitable sensing layer, the desired detection limit, the sample matrix, and available resources.

Calixarenes have been extensively studied for their versatile properties as stable building blocks upon which functional groups can be fine-tuned to form supramolecular complexes with diverse analytes in both aqueous and organic solutions [24]. Zeybek et al. reported the use of a QCM sensor coated with a lower-rim substituted diamide calix[4]arene for the detection of gas-phase benzene, toluene and *m*-xylene [25]. We previously employed a lower- and upper-rim-functionalized calix[4]arene methoxy ester (CME) as a sensing layer on gold (Au)-coated microcantilevers to detect metal ions and their counterions in aqueous solutions [26,27]. The study presented herein extends the application of the same calixarene as a sensing layer on Au-coated QTFs to detect the aromatic hydrocarbons benzene, toluene, and ethylbenzene in dilute water solutions. The experimental results were augmented with a quantum chemical density functional theory (DFT) study to provide an insight into the nature of the hypothetical interactions between the individual aromatic hydrocarbons and the CME receptor molecule.

2. Results

2.1. Resonance Frequency Measurements

Resonance frequency measurements obtained with Au-coated QTFs functionalized with THE calix[4]arene methoxy ester sensing layer (Figure 1) were used to quantitatively and qualitatively detect the presence of benzene, toluene, and ethylbenzene in water solutions. The resonance frequency of the QTF was scanned from 30.9 to 35.0 kHz using the software control of the system previously described by us [22]. Each cycle was completed in a maximum of approximately 30 s. The resonance frequency shifts were calculated using Equation (1) [22,23], and the results are shown in Figure 1d.

$$\Delta f = f_{ref} - f_a \tag{1}$$

where f_{ref} and f_a are the resonance frequencies of the distilled water control and the tested analytes at the 10^{-12}, 10^{-10}, 10^{-8}, and 10^{-6} M concentrations.

Figure 1a–d show the comparison of the resonance frequency responses of the coated QTFs when immersed in the different concentrations (10^{-12} to 10^{-6} M) of the individual analytes benzene, toluene, and ethylbenzene in deionized water (DI) solutions. Figure 1a shows that the largest resonance frequency shift, Δf = 198 Hz, is the decrease of the resonance frequencies from 32,494 Hz to 32,296 Hz measured after immersion in the 10^{-6} M solution of benzene in DI. Smaller, linear shifts were recorded for the more-diluted solutions. Resonance frequency decreases were also observed for the other analyte solutions, as shown in Figure 1b,c and a detailed discussion follows in Section 3. The highest resonance frequency shift for the corresponding 10^{-6} M solutions was in the following order: benzene (198 Hz) > toluene (191 Hz) > ethylbenzene (149 Hz). The responses of the Au-coated QTFs without the CME SAMs showed very small frequency shifts in the following order: benzene (35 Hz) > toluene (34 Hz) > ethylbenzene (31 Hz), as shown in Figure S3.

2.2. Quantum Chemical DFT Calculations

To gain a better understanding of the interaction between the CME with each of the three analytes, quantum chemical density functional theory (DFT) calculations were conducted using *Gaussian 16, Revision C.01* [28], with the CAM-B3LYP [29] and PBE [30] functionals with the LANL2DZ basis set [31], in the gas phase and water solvent system. Furthermore, the B3LYP-D3/GenECP and LanL2DZ [31] basis sets were used for Au, and the 6-311++g(d,p) basis set was used for all other atoms. The electronic interaction energy (ΔE_{int} kJ mol^{-1}) values were calculated using Equations (2) and (3) for the components of the modeled hypothetical 1:1 supramolecular complexes formed by the Au-bonded receptor CME with the respective analytes. These results are discussed in further detail in the Section 3.2.

$$\Delta E_{int} \text{ for CME} \supset \text{Analyte} = E_{[CME] \supset [Analyte]} - (E_{[CME]} + E_{[Analyte]}) \tag{2}$$

$$\Delta E_{int} \text{ for [Au4-CME]} \supset \text{Analyte} = E_{[Au4-CME] \supset [Analyte]} - (E_{[CME]} + E_{[Analyte]}) \tag{3}$$

where $E_{[CME] \supset [Anayte]}$ = optimized energy of the CME complex(es) with the specific analyte; $E_{[Au4-CME] \supset [Analyte]}$ = optimized energy of the Au4-CME 1:1 supramolecular complex(es) with the specific analyte; $E_{[Au4-CME]}$ = optimized electronic energy of the Au-receptor Au4-CME; $E_{[CME]}$ = optimized electronic energy of the free receptor CME; $E_{[Analyte]}$ = optimized electronic energy of the specific benzene, toluene. or ethylbenzene analyte.

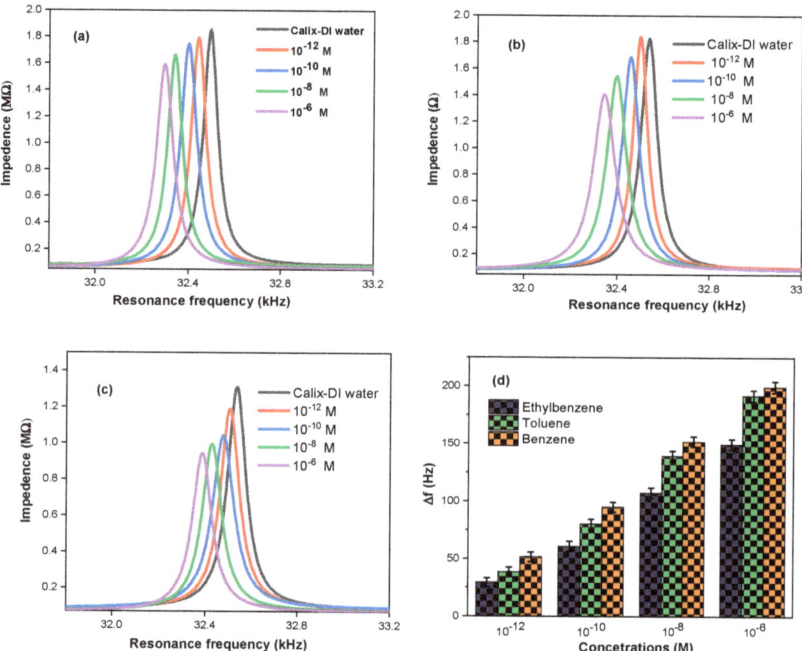

Figure 1. Comparison of the resonance frequency responses of the CME-functionalized Au-coated QTFs with the different concentrations (10^{-12} M to 10^{-6} M) of each of the aqueous solutions of (**a**) benzene, (**b**) toluene, and (c) ethylbenzene; (**d**) is a summary histogram showing the relative resonance frequency shifts (Δf) +/− ~5% of the three aromatic hydrocarbons tested.

The DFT-calculated electronic binding interaction energies (ΔIE kJ mol^{-1}) are summarized in Tables 1–3. The negative ΔIE values represent stronger interactions between the CME receptor with the analytes, which correlates with our experimental results. Table 1 shows the electronic binding interaction energies (ΔIE kJ mol^{-1}) with uncorrected BSSE values with the CAM-B3LYP/LAN2DZ and PBE/LANL2DZ functionals in the water solvent system. The calculated gas phase and water solvent system ΔIEs with the CAM-B3LYP/LAN2DZ and PBE/LANL2DZ functionals were in the following order: [CME⊃[Benzene] > [CME]⊃[Toluene] > [CME]⊃[Ethylbenzene] −51.99, −45.66, and −41.92 kJ mol^{-1}, respectively, in the gas phase and −32.12, −31.28, and −30.5 kJ mol^{-1}, respectively, in the water solvent system. A similar trend was also observed with PBE/LANL2DZ in the gas phase and the water solvent system. The results from both methods were consistent. The DFT-calculated results strongly supported the experimentally obtained results.

Table 1. Comparison of DFT-calculated electronic binding interaction energies (ΔIE kJ mol^{-1}) for 1:1 complexes of the receptor CME with benzene, toluene, and ethylbenzene with the CAM-B3LYP/LANL2DZ and PBE/LANL2DZ functionals/basis sets, in the gas phase.

| Complex | ΔIEs (kJ mol^{-1}) of the Hypothetical 1:1 Supramolecular Structures of CME with Benzene, Toluene, and Ethylbenzene | | | |
| | CAM-B3LYP/LANL2DZ | | PBE/LANL2DZ | |
	Uncorrected BSSE	Corrected BSSE	Uncorrected BSSE	Corrected BSSE
[CME]⊃[Benzene]	−51.99	−34.19	−48.18	−31.08
[CME]⊃[Toluene]	−45.66	−30.66	−44.91	−29.95
[CME]⊃[Ethylbenzene]	−41.92	−25.06	−43.08	−26.13
[Au4-CME]⊃[Benzene]	−71.42	−48.60	−61.53	−44.18
[Au4-CME]⊃[Toluene]	−65.40	−44.85	−59.40	−42.57
[Au4-CME]⊃[Ethylbenzene]	−54.39	−42.38	−51.57	−41.14

Table 2. Comparison of DFT-calculated electronic binding interaction energies (ΔIE kJ mol^{-1}) for 1:1 complexes of CME with benzene, toluene, and ethylbenzene with the CAM-B3LYP/LANL2DZ and PBE/LANL2DZ functionals/basis sets, in water solvent system.

| Complex | ΔIEs (kJ mol^{-1}) of the Hypothetical 1:1 Supramolecular Structures of CME with Benzene, Toluene, and Ethylbenzene | |
| | CAM-B3LYP/LANL2DZ | PBE/LANL2DZ |
	Uncorrected BSSE	Uncorrected BSSE
[CME]⊃[Benzene]	−32.12	−29.77
[CME]⊃[Toluene]	−31.28	−28.80
[CME]⊃[Ethylbenzene]	−30.55	−28.34
[Au4-CME]⊃[Benzene]	−60.45	−53.31
[Au4-CME]⊃[Toluene]	−58.06	−50.93
[Au4-CME]⊃[Ethylbenzene]	−55.47	−47.19

Table 3. Comparison of DFT-calculated electronic binding interaction energies (ΔIE kJ mol^{-1}) for the hypothetical 1:1 supramolecular complexes of Au4-CME with benzene, toluene, and ethylbenzene using B3LYP-D3/GenECP and LanL2DZ basis sets for Au and the 6-311++g(d,p) basis set for all other atoms in water solvent system.

| Complex | ΔIEs (kJ mol^{-1}) of the Hypothetical 1:1 Supramolecular Structures of CME with Benzene, Toluene, and Ethylbenzene | | |
| | Gas Phase | | Water Solvent |
	Uncorrected BSSE	Corrected BSSE	Uncorrected BSSE
[Au4-CME]⊃[Benzene]	−96.86	−86.39	−84.98
[Au4-CME]⊃[Toluene]	−87.80	−77.23	−75.25
[Au4-CME]⊃[Ethylbenzene]	−79.33	−67.63	−66.28

3. Discussion

3.1. Analysis of the Resonance Frequency Measurements

The Au-coated QTFs were functionalized with CME as previously described (Figure 1). The resonance frequency changes that were measured at increasing concentrations of the analytes in DI showed that the methodology was sufficiently sensitive to detect concentrations of the three aqueous analyte solutions as low as 10^{-12} M. All experiments were conducted in triplicate, separately, at four different concentrations (10^{-12}, 10^{-10}, 10^{-8}, and 10^{-6} M) for each of the analytes, benzene, toluene, and ethylbenzene. All replicates were within SD of $+/-$ ~5%. QTF sensor measurements were performed between each solute measurement with Milli-Q DI water as control experiments. Furthermore, measurements were conducted using Au-coated QTFs without any of the sensing CME to confirm that there were no frequency changes, thus indicating that no interactions of the analytes occurred with the Au-coated QTFs' surfaces alone. Figure 1a shows the largest resonance frequency shift, Δf = 199 Hz, which was measured with the CME-functionalized QTF in aqueous 10^{-6} M benzene solution. Figure 1b,c show the corresponding largest resonance frequency shifts of Δf = 191 Hz and 149 Hz, for the aqueous 10^{-6} M toluene and aqueous ethylbenzene solutions, respectively. Figure 1d illustrates the resonance frequency shifts for the corresponding 10^{-6} to 10^{-12} M solutions of the three analytes, and in all cases, the order was as follows: benzene > toluene > ethylbenzene. Their limits of detection (LODs) [32–34] were 3.47×10^{-14} M, 9.43×10^{-14} M, and 1.75×10^{-13} M, respectively, as calculated from their respective linear fit curves by plotting the logarithm concentration versus the frequency shift using Equation (4) and shown in Figure 2.

$$\mathrm{Log}\,(\mathrm{LOD}) = \left(\frac{3.3\,\sigma}{m}\right) \tag{4}$$

where σ = standard deviation of the intercept and m = slope of the fit curve.

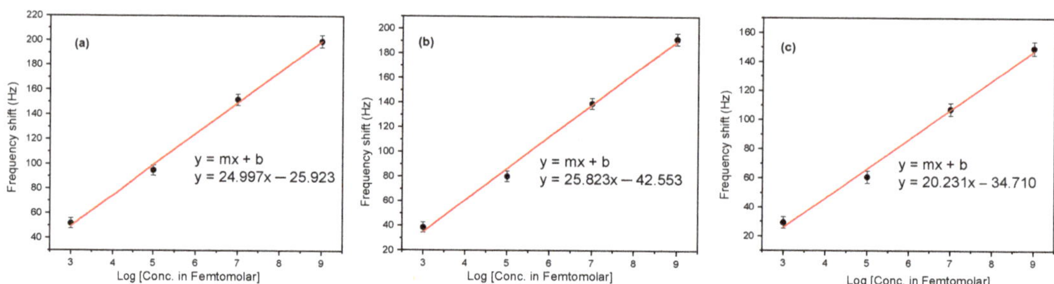

Figure 2. Comparison of the LODs of the CME-functionalized Au-coated QTFs with the different concentrations (10^{-12} M to 10^{-6} M) of each of the aqueous aromatic hydrocarbon solutions: (**a**) benzene, (**b**) toluene, and (**c**) ethylbenzene. These black dots reflect error bars of approx 5%.

3.2. DFT Calculations

The initial molecular structures of all of the examined structures were drawn using *GaussView 6.0.16* [35], and all computations were conducted using *Gaussian 16, Revision C.01* [28]. Vibrational frequency analyses were conducted for each optimized structure to ensure that they each had a vibrational minimum energy and no imaginary frequencies. We hypothesized that by immersing a Au-coated QTF into the CME solution (at a concentration of 1.0×10^{-6} M) and then incubating it for 1 h, self-assembled monolayers (SAMs) of the CME would form on the gold surfaces of the QTFs. This would be due to the formation of the covalent S-Au bond and the acetyl (CH_3C=O) fragments, as we previously showed [22,27,36,37]. A schematical illustration is shown on the left side in Figure 3. The methoxy ester group of the calixarene was further hypothesized to form supramolecular 1:1 complexes with each of the analytes, by electrostatic noncovalent "host–guest" interactions, as depicted in Figure 3a–d. To simplify the quantum chemical calculations and reduce the computation time, DFT calculations with the CAM-B3LYP/LANL2DZ and PBE/LANL2DZ functionals were also conducted using a single molecule of the CME bonded onto a gold cluster (Au4) for the complexation, as shown in Figure 3e–h, with the analytes, first in the gas phase and, then, also, in the water solvent system. Furthermore, the geometries of the hypothetical Au4-CME ([Au4-CME]) structures as their complexes with the analytes were conducted with B3LYPD3/GenECP and two different basis sets; the 6-311++g(d,p) basis set was used for the C, O, S, and H atoms (Equation (5)), and the LANL2DZ basis set was used for the Au atoms (Equation (6)), in both the gas phase and water solvent system. In order to access both basis set functions, the GenECP keyword was used. The Los Alamos National Laboratory 2 double ζ (LANL2DZ) effective-core-potential (ECP)-type basis set has been employed for heavy metal atoms, as is widely accepted [38], so employing the LANL2DZ for the gold (Au) was effective at reducing the computational resources. To avoid basis set superposition errors (BSSEs) [39–42] in the non-covalent interactions between receptor molecules and analytes, counterpoise (CP) corrections were made in the gas phase at the same level of theory. The binding interaction energies were calculated using the following equations:

$$\Delta E_{BSSE} \ (CME \supset Analyte) = E_{([CME] \supset [Analyte])} - E_{([CME])} - E_{([Analyte])} \tag{5}$$

$$\Delta E_{BSSE} \ ([Au4\text{-}CME] \supset Analyte) = E_{([Au4\text{-}CME] \supset [Analyte])} - E_{([Au4\text{-}CME])} - E_{([Analyte])} \tag{6}$$

where $E_{([CME] \supset [Analyte])}$ = optimized energy of the CME complex(es) with the specific analyte; $E_{([Au4\text{-}CME] \supset [Analyte])}$ = optimized energy of the receptor Au4-CME 1:1 complex(es) with each of the analytes; $E_{([Au4\text{-}CME])}$ = optimized electronic energy of the free receptor Au4-CME; $E_{([CME])}$ = optimized electronic energy of the free receptor CME; $E_{([Analyte])}$ = optimized electronic energy of the specific benzene, toluene, or ethylbenzene analyte.

The DFT-calculated electronic binding interaction energies (ΔIE kJ mol^{-1}) are summarized in Tables 1–3. Table 1 shows the gas phase uncorrected BSSE and corrected BSSE electronic ΔIE values using CAM-B3LYP/LANL2DZ and PBE/LANL2DZ. The ΔIEs for the hypothetical 1:1 [Au4-CME]$\supset$analyte complexes were higher than the corresponding [CME$\supset$analyte] complexes. The 1:1 [Au4-CME]$\supset$[benzene], [Au4-CME]$\supset$[toluene] and [Au4-CME]$\supset$[ethylbenzene] complexes had the highest BSSE-uncorrected ΔIE values, -71.42, -65.40, and -54.39 kJ mol^{-1}, respectively, and -48.60, -44.85, and -42.38 kJ mol^{-1}, respectively, for the respective BSSE-corrected values with CAM-B3LYP/LANL2DZ in the gas phase. The same trends, albeit of lower energies, were also observed with PBE/LANL2DZ in the gas phase. Table 2 shows similar trends as for Table 1 for the electronic binding interaction energies with the CAM-B3LYP/LANL2DZ and PBE/LANL2DZ functionals and basis sets, in the water solvent system. The system with the former gave better results compared with PBE/LANL2DZ for both the gas phase and water solvent system.

Table 3 shows the computed data for the gas phase and water solvent system for the [Au4-CME]$\supset$[benzene], [Au4-CME]$\supset$[toluene], and [Au4-CME]$\supset$[ethylbenzene] 1:1 complexes using B3LYP-D3/GenECP with LanL2DZ for Au and 6-311++g(d,p) for all other atoms. The highest BSSE-uncorrected ΔIE values were obtained with this functional/basis set: -96.86, -87.80, and -79.33 kJ mol^{-1}, respectively, and -86.39, -77.23, and -67.63 kJ mol^{-1}, respectively, for the corresponding BSSE-corrected values in the gas phase. The highest BSSE-uncorrected ΔIE values were -84.98, -75.25, and -66.28 kJ mol^{-1} in the water solvent. The DFT-calculated results strongly supported the experimentally obtained results, and although the same trends were seen with either functionals or basis sets, the B3LYP-D3/GenECP afforded the best results, i.e., gave the most-energetically favored electronic binding interaction energies. Uddin et al. reported the use of both CAM-B3LYP/LAN and PBE/LANL2DZ along with other levels of theory to identify variations with functionals and basis sets [43].

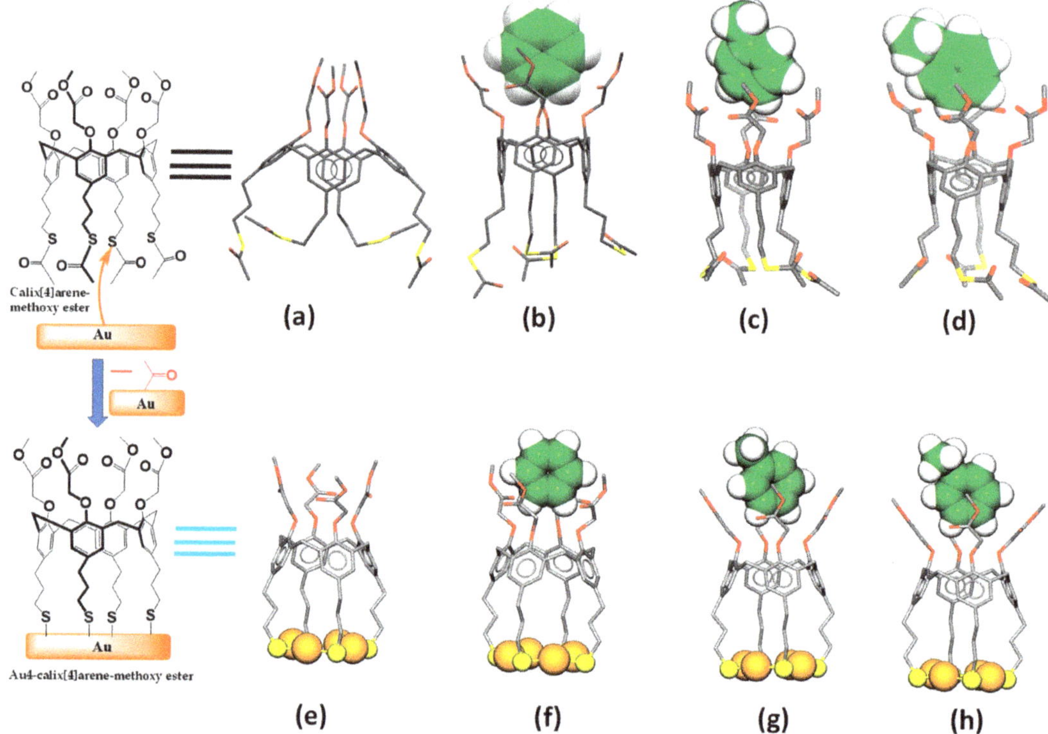

Figure 3. Geometry-optimized structures: (**a**) receptor molecule CME; (**e**) receptor Au4-CME; 1:1 binding modes of (**b**) CME⊃benzene, (**c**) CME⊃toluene, (**d**) CME⊃ethylbenzene, (**f**) Au4-CME⊃benzene, (**g**) Au4-CME⊃toluene, and (**h**) Au4-CME⊃ethylbenzene. Color code: carbon = gray (except benzene, toluene, and ethylbenzene carbon = green); gold = orange; oxygen = red; sulfur = yellow; hydrogen atoms = white for benzene, toluene, and ethylbenzene (other hydrogen atoms are omitted for clarity).

To gain additional understanding of the nature of the interactions and stabilities between the free and Au-bound CME molecules with the three analytes, frontier molecular orbitals (FMOs) [44] were examined. These were based on the most-stable geometries generated for these molecules and their respective highest occupied molecular orbitals (HOMOs) and lowest unoccupied molecular orbitals (LUMOs), as shown in Figure 4.

Koopmans' theorem considers that the energy levels of the HOMOs and LUMOs are related to their nucleophilic and electrophilic properties, respectively [45]. Large HOMO–LUMO gaps are indicative of the chemical species' high stability and low reactivity. Table 4 shows that the HOMO–LUMO value for Au4-CME in the gas phase (7.050 eV) was lower than the water solvent system (7.319 eV). The lower HOMO–LUMO gap value of the receptor Au4-CME, therefore, understandably had a higher binding ability with the analytes in the gas phase compared with the water solvent system and the DFT-calculated binding energies as more negative for the gas phase. Presumably, the solvation of each molecule in the water solvent prevented their effective interactions. The solvation energy in water is more complex with hydrogen bonding having significant effects on the solvation energies in water. The HOMO–LUMO energy gaps for the complexes were in the following order: [Au4-CME⊃[benzene] > [Au4-CME]⊃[toluene] > [Au4-CME]⊃[ethylbenzene]. The HOMO–LUMO energy values for the CME, Au4-CME, each of the analytes, and their complexes are reported in Tables S1 and S2 (Supporting Information) and can be used to calculate a number of other significant and valuable quantum chemical properties such as the global hardness (η), global softness (S), electrophilicity index (ω), electronegativity (χ), and chemical potential (μ), which all measure chemical reactivity. These were generated using *GaussView 6.0*.16 [35]. The chemical potential (μ), hardness (η), softness (S), and global electrophilicity index (ω) of the receptor molecules CME, Au4-CME, analytes, and their complexes were calculated using the HOMO and LUMO energies. The formulas used for these calculations are given as follows: $\eta = (E_{LUMO} - E_{HOMO})/2$; $\mu = (E_{HOMO} - E_{LUMO})/2$; $S = 1/\eta$ and $\omega = \mu^2/2\eta$.

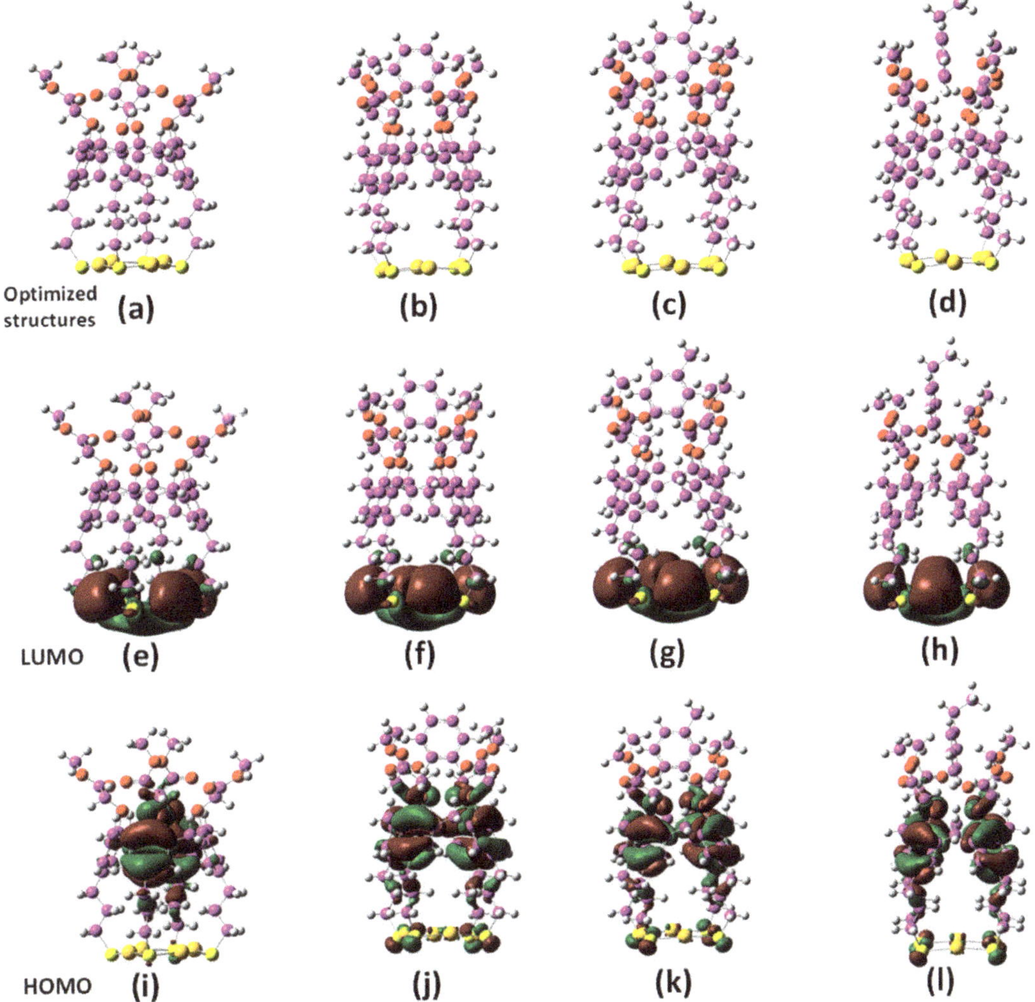

Figure 4. Geometry-optimized structures of Au4-CME and its 1:1 supramolecular complexes with benzene, toluene, and ethylbenzene (**a–d**) and their respective HOMO (**e–h**) and LUMO (**i–l**) FMOs. Color code: carbon = purple; gold = orange; oxygen = red; sulfur = yellow; hydrogen atoms = white.

Table 4. DFT-calculated global scalar properties of the receptor molecule Au4-CME and its complexes with the analytes at the CAM-B3LYP/LANL2DZ level of theory in the gas phase and water solvent system.

	HOMO Energy eV	LUMO Energy eV	H–L Gap eV	Ionization Potential (IP) eV	Electron Affinity (EA) eV	Electronegativity (χ) eV	Chemical Potential (μ) eV	Hardness (η) eV	Softness (S) eV	Electrophilicity Index (ω) eV
Gas phase										
Au4-Calix[4]arene	−7.230	−0.181	7.050	7.230	0.181	3.706	−3.706	3.525	0.284	1.948
Au4-CME⊃benzene	−7.337	−0.202	7.135	7.337	0.202	3.770	−3.770	3.568	0.280	1.992
Au4-CME⊃toluene	−7.325	−0.196	7.129	7.325	0.196	3.761	−3.761	3.564	0.281	1.984
Au4-CME⊃ethylbenzene	−7.321	−0.195	7.126	7.321	0.195	3.758	−3.758	3.563	0.281	1.982
Water solvent system										
Au4-CME	−7.526	−0.206	7.319	7.526	0.206	3.866	−3.866	3.660	0.273	2.042
Au4-CME⊃benzene	−7.625	−0.160	7.465	7.625	0.160	3.892	−3.892	3.732	0.268	2.030
Au4-CME⊃toluene	−7.622	−0.159	7.463	7.622	0.159	3.890	−3.890	3.732	0.268	2.028
Au4-CME⊃ethylbenzene	−7.605	−0.162	7.443	7.605	0.162	3.883	−3.883	3.722	0.269	2.026

4. Materials and Methods

4.1. Chemicals and Materials

Benzene, toluene, ethylbenzene, dichloromethane, and ethanol were procured from Sigma-Aldrich, St. Louis, MI, USA. The synthesis of the receptor sensing CME was previously reported by us [27]. All aqueous solutions of the aromatic compounds were prepared as saturated solutions using deionized (DI) water with a resistivity of 18.6 MΩ·cm. The pH of the deionized water was 6.82. The analyte solutions had a pH of 6.85. The Au-coated QTFs were purchased from Forien Inc., Edmonton, AB, Canada (https://www.fourien.com/). They were coated with Au using a vacuum evaporation method; the thickness of the gold coating was around 100 nm. The resonance frequency of the QTFs was 32.768 kHz; the spring constant was ~20 kN/m; the load capacitance was 12.5 pF. A 10^{-6} M solution of the CME was used to functionalize the Au-coated QTFs.

4.2. Experimental Setup and Instrumentation

The frequencies of QTF resonance were measured with a Quester Q10 instrument made by Fourien Inc. The Quester Q10 contains an impedance analyzer that performs frequency sweeps and measures the impedance response's real and imaginary components. To resonate the QTFs at specific frequencies, the proportional–integral–differential technique was used in this system. The system can keep the QTF at a fixed distance from any analyte solute when combined with a translation stage. The data collected were analyzed using MATLAB or the Origin Lab program. The system is illustrated in Figure 5, which includes a schematic of the methoxy ester CME SAM-functionalized Au-coated QTFs. A detailed description of the instrumental setup and software integration can be found in our previous study and elsewhere [22,23].

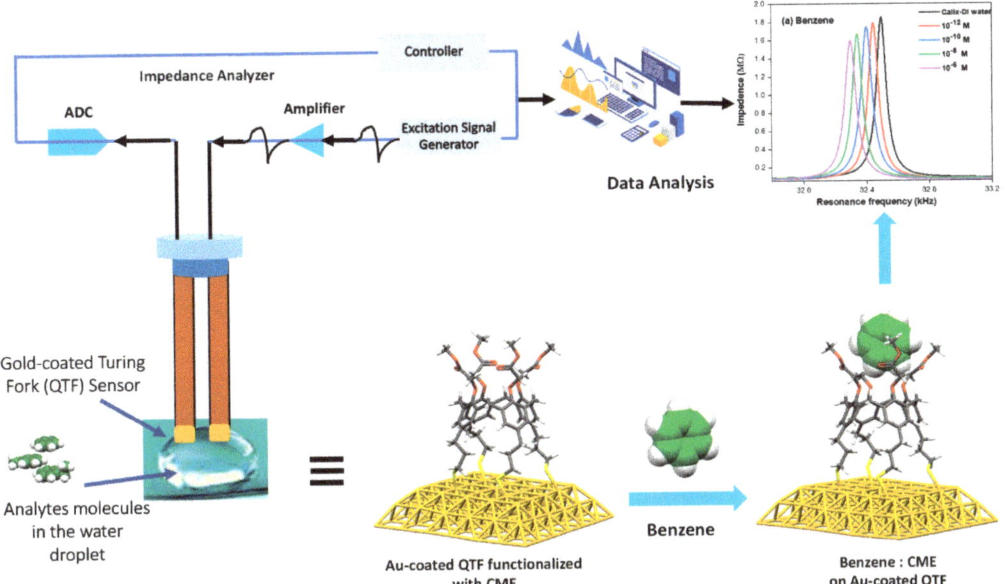

Figure 5. Schematical illustration of the QTF measurement system used and the Au-coated QTFs functionalized with CME and its hypothetical 1:1 supramolecular complex with benzene; (a) is a typical output showing the resonance frequency responses (in kHz) to the aqueous benzene solutions of different concentrations (10^{-6}–10^{-12} M). Color code: carbon = gray (except benzene carbon = green); hydrogen = white; sulfur = yellow; oxygen = red; hydrogen atoms = white.

4.3. Experimental Methods

The gold-coated QTFs self-assembled monolayers (SAMs) of calix[4]arene were prepared by incubating the QTFs for 1 h in a solution of CME (1.0×10^{-6} M) at room temperature in a 1:9 dichloromethane:ethanol solvent. The QTFs were then washed three times with the same solvent mixture to remove any unbonded CME and were then dried using a nitrogen gas stream. The resonance frequency of the functionalized QTFs was measured in DI water for reference. Finally, the functionalized QTFs were exposed to different concentrations (10^{-12}, 10^{-10}, 10^{-8}, and 10^{-6} M) of saturated aqueous (DI) solutions of the analytes (benzene, toluene, and ethylbenzene) for ten minutes, after which time, their resonance frequencies were measured to determine their sensitivity. To maintain consistent experimental conditions, the QTFs were directly immersed at a depth of 25 μm in a small droplet of a 100 μL volume of the respective analyte solutions.

5. Conclusions

In this study, we investigated the effectiveness of using a CME as a sensing receptor molecule for forming stable SAMs on the gold surfaces of QTFs. Our findings suggested that these SAMs have the potential for rapidly detecting the aromatic hydrocarbons benzene, toluene, and ethylbenzene in water samples, at very low analyte concentrations (10^{-12} M) and using small amounts of aqueous solutions (100 μL). Among the analytes that were tested, benzene had the highest effect on the QTF responses, with frequency shifts in the following order: benzene (199 Hz) > toluene (191 Hz) > ethylbenzene (149 Hz). This trend is consistent with the trends observed by Zeybek et al. for the vapor-phase BTX detections using a QCM with a thin-film layer of a calix[4]arene with a different functional group [25].

In our study, the limits of detection for benzene, toluene, and ethylbenzene were calculated to be 3.5×10^{-14} M, 9.4×10^{-14} M, and 1.8×10^{-13} M, respectively. Utilizing density functional theory (DFT), the interaction energies (ΔIEs) were calculated for the hypothetical 1:1 supramolecular complexes formed between the analytes within the methoxy ester moieties of the gold-bonded CME. The DFT-calculated ΔIE values for the 1:1 supramolecular analyte complexes with the Au-bonded Au4-CME receptor showed higher values compared to the free CME. Among the density functionals we tested, the B3LYPD3/GenECP functional with the effective core potential LANL2DZ for Au atoms and the 6-311++g(d,p) basis set for C, H, O, and S atoms showed the optimal binding interactions of the receptor CME with the analytes. The complexes of [Au4-CME]⊃[benzene], [Au4-CME]⊃[toluene], and [Au4-CME]⊃[ethylbenzene] had the highest BSSE-uncorrected and BSSE-corrected binding interaction (ΔIE) values, which were −96.86, 87.80, and −79.33 kJ mol^{-1}, respectively, and −86.39, −77.23, and −67.63 kJ mol^{-1}, respectively, for the respective BSSE-corrected values at the B3LYPD3/GenECP with the LANL2dZ and 6311++g(d,p) level of theory in the gas phase. Similar trends were observed when considering the hypothetical Au-CME complexes with the analytes, using both the CAM-B3LYP/LANL2DZ and LANL2DZ/PBE0 levels of theory in the gas phase and water solvent system. The DFT results showed that there was consistency in the ΔIE values obtained from the three above-mentioned methods. The ΔIE values of the selected DFT functionals along with the LANL2DZ and the 6-311++g(d,p) basis sets were in the following order: B3LYPD3 > CAM-B3LYP > PBE0. The results obtained through DFT calculations, considering both uncorrected and corrected BSSE values, strongly supported the experimental findings on the proposed binding modes between the analytes (benzene, toluene, and ethylbenzene) and the CME SAMs on the Au-QTF sensing surfaces.

Supplementary Materials: The following Supporting Information can be downloaded at: https://www.mdpi.com/article/10.3390/molecules28196808/s1, Figure S1. Geometry-optimized structures: (a) Receptor molecule CME; (e) Receptor Au4-CME; 1:1 binding modes of (b) CME⊃benzene; (c) CME⊃toluene; (d) CME⊃ethylbenzene; (f) Au4-CME⊃benzene; (g) Au4-CME⊃toluene; and (h) Au4-CME⊃ethylbenzene; Figure S2. Comparison of the resonance frequency responses of the Au-coated QTFs (without functionalized calix[4]arene) with the different concentrations (10^{-12} M to 10^{-6} M) of each of the aqueous of aromatic hydrocarbons solution; Table S1. Calculated global scalar properties of benzene, toluene, ethylbenzene, calix[4]arene methoxy ester, Au4-calix[4]arene methoxy ester and it's 1:1 complex in the gas phase; Table S2. Calculated global scalar properties of benzene, toluene, ethylbenzene, calix[4]arene methoxy ester, Au4-calix[4]arene methoxy ester and it's 1:1 complex with analytes (benzene, toluene and ethylbenzene) at the CAM-B3LYP/LANL2DZ level of theory in water solvent system; Table S3. Calculated global scalar properties of benzene, toluene, ethylbenzene, calix[4]arene methoxy ester, Au4-calix[4]arene methoxy ester and it's 1:1 complex with analytes (benzene, toluene and ethylbenzene) at the PBE0/LANL2DZ level of theory in the gas phase; Table S4. Calculated global scalar properties of benzene, toluene, ethylbenzene, calix[4]arene methoxy ester, Au4-calix[4]arene methoxy ester and it's 1:1 complex with analytes (benzene, toluene and ethylbenzene) at the PBE0/LANL2DZ level of theory in water solvent system.

Author Contributions: Conceptualization, S.R., A.N.A., P.E.G. and M.A.A.-G.; methodology, S.R., A.N.A. and M.A.A.-G.; validation, S.R., H.A. and A.N.A.; investigation, F.S.A., W.K.A. and N.A.; data analysis, F.S.A., W.K.A. and N.A.; writing—original draft preparation, S.R., A.N.A., P.E.G. and M.A.A.-G.; writing—review and editing, S.R., A.N.A., H.A., M.A.A.-G. and P.E.G.; supervision reviewing and editing, S.R., A.N.A. and P.E.G.; software, S.R. and M.A.A.-G.; funding acquisition, S.R. All authors have read and agreed to the published version of the manuscript.

Funding: The authors extend their appreciation to the Deputyship for Research & Innovation, "Ministry of Education" in Saudi Arabia for funding this research (IFKSUOR3–303-1).

Institutional Review Board Statement: Not applicable.

Informed Consent Statement: Not applicable.

Data Availability Statement: Any additional data in support of the findings of this study besides those provided as Supplementary Materials are available from the corresponding author upon reasonable request.

Acknowledgments: The authors extend their appreciation to the Deputyship for Research & Innovation, "Ministry of Education" in Saudi Arabia for funding this research (IFKSUOR3–303-1). The DFT and quantum chemical computations were enabled by the support in part provided by Oliver Stueker at Acenet and the Digital Research Alliance of Canada, who are gratefully appreciated.

Conflicts of Interest: The authors declare no conflict of interest.

Sample Availability: Not applicable.

References

1. Chary, N.S.; Fernandez-Alba, A.R. Determination of volatile organic compounds in drinking and environmental waters. *Trends Anal. Chem.* **2012**, *32*, 60–75. [CrossRef]
2. Yu, B.; Yuan, Z.; Yu, Z.; Xue-song, F. BTEX in the environment: An update on sources, fate, distribution, pretreatment, analysis, and removal techniques. *Chem. Eng. J.* **2022**, *435*, 134825. [CrossRef]
3. Chen, D.; Lawrence, K.G.; Sandler, D.P. Nontraditional occupational exposures to crude oil combustion disasters and respiratory disease risk: A narrative review of the literature. *Curr. Allergy Asthma Rep.* **2023**, *23*, 299–311. [CrossRef] [PubMed]
4. Nriagu, J.; Udofia, E.A.; Ekong, I.; Ebuk, G. Health risks associated with oil pollution in the Niger Delta, Nigeria. *Int. J. Environ. Res. Public Health* **2016**, *13*, 346. [CrossRef] [PubMed]
5. Ramírez, N.; Cuadras, A.; Rovira, E.; Borrull, F.; Marcé, R.M. Chronic Risk Assessment of Exposure to Volatile Organic Compounds In The Atmosphere Near The Largest Mediterranean Industrial Site. *Environ. Int.* **2012**, *39*, 200–209. [CrossRef] [PubMed]
6. Honda, M.; Suzuki, N. Toxicities of Polycyclic Aromatic Hydrocarbons for Aquatic Animals. *Int. J. Environ. Res. Public Health.* **2020**, *17*, 1363. [CrossRef] [PubMed]
7. Rajabi, H.; Mosleh, M.H.; Mandal, P.; Lea-Langton, A.; Sedighi, M. Emissions of volatile organic compounds from crude oil processing–Global emission inventory and environmental release. *Sci. Total Environ.* **2020**, *727*, 138654. [CrossRef] [PubMed]
8. Available online: https://pubchem.ncbi.nlm.nih.gov/compound/241 (accessed on 15 September 2023).

9. Yang, Y.; Miller, D.J.; Hawthorne, S.B. Toluene solubility in water and organic partitioning from gasoline and diesel fuel into water at elevated temperatures and pressures. *J. Chem. Eng. Data* **1997**, *42*, 908–913. [CrossRef]

10. Available online: https://pubchem.ncbi.nlm.nih.gov/compound/7500 (accessed on 15 September 2023).

11. US Environmental Protection Agency. National Primary Drinking Water Regulations. Available online: https://www.epa.gov/ground-water-and-drinking-water/national-primary-drinking-water-regulations (accessed on 29 August 2023).

12. Adeniran, M.A.; Oladunjoye, M.A.; Doro, K.O. Soil and groundwater contamination by crude oil spillage: A review and implications for remediation projects in Nigeria. *Front. Environ. Sci.* **2023**, *11*, 1137496. [CrossRef]

13. Zogorski, J.S.; Carter, J.M.; Ivahnenko, T.; Lapham, W.W.; Moran, M.J.; Rowe, B.L.; Squillace, P.J.; Toccalino, P.L. Volatile organic compounds in the nation's ground water and drinking-water supply wells. *US Geol. Surv. Circ.* **2006**, *1292*, 101.

14. Smith, M.T. Advances in understanding benzene health effects and susceptibility. *Annu. Rev. Public Health* **2010**, *31*, 133–148. [CrossRef] [PubMed]

15. Lee, E.H. Iron oxide catalysts for dehydrogenation of ethylbenzene in the presence of steam. *Catal. Rev. Sci. Eng.* **1974**, *8*, 285–305. [CrossRef]

16. Liu, Q.; Liu, Y.; Zhang, M. Source Apportionment of Personal Exposure to Carbonyl Compounds and BTEX at Homes in Beijing, China. *Aerosol Air Qual. Res.* **2014**, *14*, 330–337. [CrossRef]

17. Jurdáková, H.; Kubinec, R.; Jurčišinová, M.; Krkošová, Ž.; Blaško, J.; Ostrovský, I.; Soják, L.; Berezkin, V.G. Gas chromatography analysis of benzene, toluene, ethylbenzene and xylenes using newly designed needle trap device in aqueous samples. *J. Chromatogr. A* **2008**, *1194*, 61–164. [CrossRef] [PubMed]

18. Pascale, R.; Bianco, G.; Calace, S.; Masi, S.; Mancini, I.M.; Mazzone, G.; Caniani, D. Method development and optimization for the determination of benzene, toluene, ethylbenzene and xylenes in water at trace levels by static headspace extraction coupled to gas chromatography–barrier ionization discharge detection. *J. Chromatogr. A* **2018**, *1548*, 10–18. [CrossRef] [PubMed]

19. Bahrami, A.; Ghamari, F.; Yamini, Y.; Ghorbani Shahna, F.; Koolivand, A. Ion-pair-based hollow-fiber liquid-phase microextraction combined with high-performance liquid chromatography for the simultaneous determination of urinary benzene, toluene, and styrene metabolites. *J. Sep. Sci.* **2018**, *41*, 501–508. [CrossRef] [PubMed]

20. Karlitschek, P.; Lewitzka, F.; Bünting, U.; Niederkrüger, M.; Marowsky, G. Detection of aromatic pollutants in the environment by using UV-laser-induced fluorescence. *Appl. Phys. B* **1998**, *67*, 497–504. [CrossRef]

21. Iyer, A.; Mitevska, V.; Samuelson, J.; Campbell, S.; Bhethanabotla, V.R. Polymer–plasticizer coatings for BTEX detection using quartz crystal microbalance. *Sensors* **2021**, *21*, 5667. [CrossRef]

22. Rahman, S.; Al-Gawati, M.A.; Alfaifi, F.S.; Muthuramamoorthy, M.; Alanazi, A.F.; Albrithen, H.; Alzahrani, K.E.; Assaifan, A.K.; Alodhayb, A.N.; Georghiou, P.E. The Effect of Counterions on the Detection of Cu^{2+} Ions in Aqueous Solutions Using Quartz Tuning Fork (QTF) Sensors Modified with L-Cysteine Self-Assembled Monolayers: Experimental and Quantum Chemical DFT Study. *Chemosensors* **2023**, *11*, 88. [CrossRef]

23. Alshammari, A.; Aldosari, F.; Qarmalah, N.B.; Lsloum, A.; Muthuramamoorthy, M.; Alodhayb, A. Detection of chemical host-guest interactions using a quartz tuning fork sensing system. *IEEE Sens. J.* **2020**, *20*, 12543–12551. [CrossRef]

24. Neri, P.; Sessler, J.L.; Wang, M.-X. (Eds.) *Calixarenes and Beyond*; Springer International Publishing AG: Cham, Switzerland, 2016.

25. Zeybek, N.; Acikbas, Y.; Bozkurt, S.; Sirit, A.; Çapan, R.; Erdoğan, M.; Özkaya, C. Developing of the calixarene based diamide chemical sensor chip for detection of aromatic hydrocarbons' vapors. *J. BAUN Inst. Sci. Technol.* **2021**, *23*, 291–300. [CrossRef]

26. Georghiou, P.E.; Rahman, S.; Assiri, Y.; Valluru, G.K.; Menelaou, M.; Alodhayb, A.N.; Braim, M.; Beaulieu, L.Y. Development of calix [4]arenes modified at their narrow- and wide-rims as potential metal ions sensor layers for microcantilever sensors: Further studies. *Can. J. Chem.* **2022**, *100*, 144–149. [CrossRef]

27. Georghiou, P.E.; Rahman, S.; Valluru, G.; Dawe, L.N.; Rahman, S.S.; Alodhayb, A.N.; Beaulieu, L.Y. Synthesis of an upper-and lower-rim functionalized calix [4] arene for detecting calcium ions using a microcantilever sensor. *New J. Chem.* **2013**, *37*, 298–1301. [CrossRef]

28. Frisch, M.J.; Trucks, G.W.; Schlegel, H.B.; Scuseria, G.E.; Robb, M.A.; Cheeseman, J.R.; Scalmani, G.; Barone, V.; Petersson, G.A.; Nakatsuji, H.; et al. *Gaussian 16, Revision C.01*; Gaussian, Inc.: Wallingford, CT, USA, 2019.

29. Yanai, T.; Tew, D.; Handy, N. A new hybrid exchange-correlation functional using the Coulomb-attenuating method (CAM-B3LYP). *Chem. Phys. Lett.* **2004**, *393*, 51–57. [CrossRef]

30. Perdew, J.P.; Burke, K.; Ernzerhof, M. Generalized gradient approximation made simple. *Phys. Rev. Lett.* **1996**, *77*, 3865. [CrossRef] [PubMed]

31. Check, C.E.; Faust, T.O.; Bailey, J.M.; Wright, B.J.; Gilbert, T.M.; Sunderlin, L.S. Addition of polarization and diffuse functions to the LANL2DZ basis set for p-block elements. *J. Phys. Chem. A* **2001**, *105*, 8111–8116. [CrossRef]

32. Armbruster, D.A.; Pry, T. Limit of blank, limit of detection and limit of quantitation. *Clin. Biochem. Rev.* **2008**, *29* (Suppl. 1), S49–S52.

33. Theodorsson, E. Limit of Blank, Limit of Detection and Limit of Quantitation. Available online: https://www.eflm.eu/files/efcc/Zagreb-Theodorsson_2.pdf (accessed on 15 September 2023).

34. Shrivastava, A.; Gupta, V.B. Methods for the determination of limit of detection and limit of quantitation of the analytical methods. *Chron. Young Sci.* **2011**, *2*, 21–25. [CrossRef]

35. Dennington, R.; Keith, T.A.; Millam, J. *GaussView, Version 6.0.16*; Semichem Inc.: Shawnee Mission, KS, USA, 2019.

36. Mandler, D.; Kraus-Ophir, S. Self-assembled monolayers(SAMs) for electrochemical sensing. *J. Solid State Electrochem.* **2011**, *15*, 1535–1558. [CrossRef]

37. Valluru, G. Synthesis and Applications of Some Upper and Lower Rim Functionalized Calix[4] Arenes and Calix [4] Naphthalene Derivatives. Ph.D. Dissertation, Memorial University of Newfoundland, St. John's, NL, Canada, August 2015.

38. Hay, P.J.; Wadt, W.R. Ab initio effective core potentials for molecular calculations. Potentials for K to Au including the outermost core orbitals. *J. Chem. Phys.* **1985**, *82*, 299–310.

39. Boys, S.; Bernardi, F. The calculation of small molecular interactions by the differences of separate total energies. Some procedures with reduced errors. *Mol. Phys.* **1970**, *19*, 553–566. [CrossRef]

40. Kestner, N.R.; Combariza, J.E. Basis set superposition errors: Theory and practice. In *Reviews in Computational Chemistry*; Wiley-VCH, John Wiley and Sons, Inc.: New York, NY, USA, 1999; Volume 13, p. 99.

41. Muhammad, S.; Irfan, A.; Al-Sehemi, A.G.; Al-Assiri, M.S.; Kalam, A.; Chaudhry, A.R. Quantum chemical investigation of spectroscopic studies and hydrogen bonding interactions between water and methoxybenzeylidene-based humidity sensor. *J. Theor. Comput. Chem.* **2015**, *14*, 1550029. [CrossRef]

42. Muhammad, S.; Liu, C.; Zhao, L.; Wu, S.; Su, Z. A theoretical investigation of intermolecular interaction of a phthalimide based "on–off" sensor with different halide ions: Tuning its efficiency and electro-optical properties. *Theor. Chem. Acc.* **2009**, *122*, 77–86. [CrossRef]

43. Uddin, K.M.; Ralph, D.; Henry, D.J. Mechanistic investigation of halopentaaquachromium (III) complexes: Comparison of computational and experimental results. *Comput. Theor. Chem.* **2015**, *1070*, 152–161. [CrossRef]

44. Fukui, K.; Yonezawa, T.; Shingu, H. A molecular orbital theory of reactivity in aromatic hydrocarbons. *J. Chem. Phys.* **1952**, *20*, 722–725. [CrossRef]

45. Luo, J.; Xue, Z.Q.; Liu, W.M.; Wu, J.L.; Yang, Z.Q. Koopmans' theorem for large molecular systems within density functional theory. *J. Phys. Chem. A* **2006**, *110*, 12005–12009. [CrossRef] [PubMed]

Article

Breaking Azacalix[4]arenes into Induline Derivatives

Zhongrui Chen, Gabriel Canard, Olivier Grauby, Benjamin Mourot and Olivier Siri *

Centre Interdisciplinaire de Nanoscience de Marseille (CINaM), UMR 7325 CNRS Aix-Marseille Université, Campus de Luminy, Case 913, F-13288 Marseille, France; zhongrui.chen@buct.edu.cn (Z.C.); gabriel.canard@univ-amu.fr (G.C.); olivier.grauby@univ-amu.fr (O.G.); benjamin.mourot@cnrs.fr (B.M.)
* Correspondence: olivier.siri@univ-amu.fr

Abstract: Tetraamino-tetranitro-azacalixarene **5** is at the crossroad of two different families of compounds depending on the conditions and the agent used to reduce the NO_2 groups: (1) azacalixphyrin **7** in neutral medium, or (2) phenazinium of type **8** in acidic medium. The key role of the N-substituted amino functions at the periphery is highlighted by investigating octaaminoazacalixarene as a model compound, and by using the corresponding tetrahydroxy-tetranitro-azacalixarene **15** as a precursor, which behaves differently.

Keywords: azacalix[4]arenes; macrocycles; phenazinium; nanostructure

Citation: Chen, Z.; Canard, G.; Grauby, O.; Mourot, B.; Siri, O. Breaking Azacalix[4]arenes into Induline Derivatives. *Molecules* **2023**, *28*, 8113. https://doi.org/10.3390/molecules28248113

Academic Editors: Roman Dembinski and Paula M. Marcos

Received: 16 November 2023
Revised: 7 December 2023
Accepted: 12 December 2023
Published: 15 December 2023

1. Introduction

Calix[n]arenes have been the focus of much attention in supramolecular chemistry for decades due to their specific molecular structure, which allows the formation of host–guest complexes [1–3]. Considerable effort is now devoted to derivatizing the basic backbones of calixarenes, and the most recent developments concern the preparation of analogues by replacing methylenic bridges with heteroatoms in order to tune and improve their properties [4]. Among the various types of heterocalixarenes, thiacalixarenes incorporating sulfur bridge atoms have been the most extensively studied, as their syntheses can be performed using simple and versatile one-step procedures [5–8]. The introduction of sulfur in place of CH_2 bridges has opened up a wide range of possibilities, offering many new features compared with "classical" calixarenes. For intance, thiacalix[4]arenes with one bridge oxidized to sulfoxide (**1**) can react with organolithium compounds to form cleaved linear structures **2** that are otherwise very difficult to access [9].

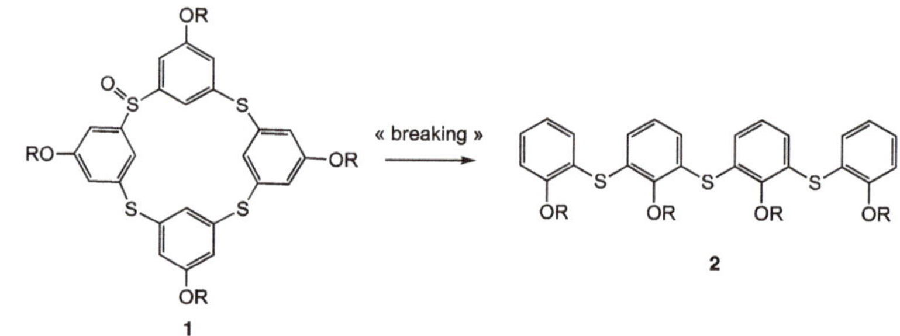

The special functionalities conferred by heteroatoms have stimulated research progress in the field of other heterocalixarenes such as oxacalixarenes [10–13] and afterwards azacalixarenes, which have been studied in less detail to date. Although the first paper on azacalixarenes **3** was reported in 1963 by Smith (X-ray structure determination) [14], it was not until the late 90s and the reported Buchwald–Hartwig amination (Pd catalyzed) that the synthesis of analogues appeared in the literature [15–21]. The introduction of nitrogen-bridging atoms on the calix[4]arene scaffold has numerous consequences such as in supramolecular chemistry with assemblies of azacalixarenes [20–22]. N-Alkylation of **3** was also a powerful tool to build more sophisticated receptors [23] or introducing inherent chirality [24]. The use of the nitrogen bridge as a spin-bearing site was on the other hand described (upon oxidation) to produce stable radical cations, high-spin diradicals, or polycationic species [25–31].

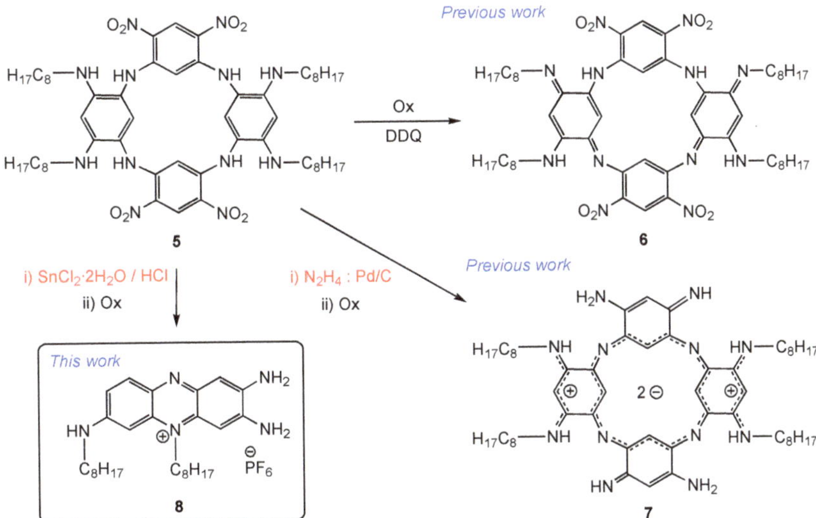

At the end of the 2000s, another approach based on aromatic nucleophilic substitutions between a 1,3-diaminobenzene derivative and an electron-poor aryl (typically 1,3-difluoro-4,6-dinitrobenzene) allowed the extending of the library of molecules (**4**) [32–34], opening up a wide range of applications [35]. One of the most striking characteristics of tetranitroazacalixarenes **4** is the strong conjugation of the nitrogen lone pair of each bridge with the NO_2 groups which is responsible for the exclusive 1,3-alternate conformation adopted by these macrocycles so far [36]. This smart and stable conformation produces a close spatial proximity between the appended substituents that were exploited in common recognition processes [34]. Next, the introduction of amino functions at the periphery of tetranitroazacalixarenes (**5**) was considered, as the macrocycle then possesses two tetraaminobenzene-type subunits, well known to be oxidized to the corresponding benzoquinonediimine due to their extremely rich electronic character [37]. This approach led to a new family of macrocycles of type **6**, called azacalixquinarenes, via an oxidation reaction with 2,3-dichloro-5,6-dicyano-1,4-benzoquinone (DDQ) (Figure 1). The resulting molecule consists of two diaminobenzoquinonediimine units linked by dinitrobenzene moieties [37].

Figure 1. Versatility of tetranitro-tetraaminoazacalixarenes **5**.

Interestingly, the electron-withdrawing nature of the dinitrobenzene moieties in **6** can trigger the intramolecular H-transfer, generating zwitterionic-ground state quinones. The nature of the N-substituents and the polarity of the solvent appear also to have a crucial impact on the equilibrium between the canonical and zwitterionic forms, which exhibit distinct optical and electrochemical properties. Beyond the interest of its oxidation, the reduction reaction of the same molecule **5** appeared also particularly attractive. Tetranitro-tetraaminoazacalixarene **5** alone is indeed remarkably specific as the reduction of its NO_2 groups with hydrogen on Pd/C leads to the formation of the corresponding, extremely unstable octaaminoazacalixarene (not isolated), which oxidizes in air to form a new class of porphyrin analogs **7** known as azacalixphyrins [38].

Herein, we wish to describe that when the same precursor **5** was instead reduced with $SnCl_2$/HCl, the formation of a pink solid, highly emissive in solution, was observed. Spectroscopic investigations demonstrated that azacalixarene **5** is in fact at the crossroad of two different families of dyes depending on the experimental conditions of reduction: the known azacalixphyrins **7** [38] or a triamino-phenazinium **8**. The key role of the amino functions at the periphery of the macrocycle was demonstrated by investigating two analogues with different functions at the periphery.

2. Results

Compound **5** was prepared as described in the literature [39]. Its reduction was carried out with $SnCl_2 \cdot 2H_2O$ (32 equiv.) in the presence of HCl (12N) under air (reflux overnight). After neutralization with $NaHCO_3$ and anion exchange reaction, **8** was obtained as a dark red solid in a 62% yield (Scheme 1). Mass spectrometry revealed a peak at m/z = 450.4 for compound **8**, reflecting a "breaking" in the macrocycle (a peak corresponding to m/z = 923.3 was expected for the macrocycle **7**). The ^1H NMR spectrum of **8** confirmed this hypothesis, notably with the presence of three aromatic protons showing a doublet signal featuring the substitution pattern indicated in Figure 2 (Ha, Hb, and Hc). Interestingly, the signals of the Hd and He protons in **8** at δ = 7.17 and 7.00 ppm, respectively, undergo a very significant shielding effect (Hd = 6.25 ppm and He = 5.91 ppm) after addition of NaOD (40 wt. % in D_2O).

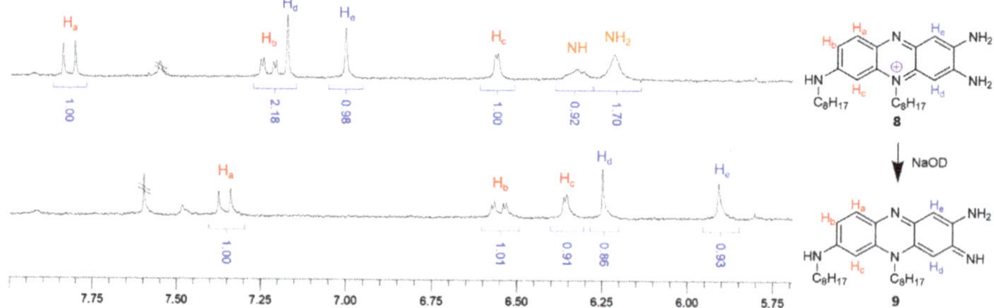

Scheme 1. Synthesis of the induline 3B derivative **8**.

Figure 2. ^1H NMR spectra of **8** (top) and **9** (bottom, after addition of NaOD) in MeCN (the window between δ = 5.7 ppm and 0 ppm has been omitted for clarity).

This observation is consistent with a quinoidal (i.e., non-aromatic) rearrangement of the molecule [40–42]. In other words, molecule **8** is capable, in basic medium, of sacrificing its aromatic character in favor of a non-aromatic species (Figure 2).

The UV-Vis absorption spectrum of phenazinium **8** in acetonitrile presents bands in the high energy range and a broad band in the green region covering the 400–600 nm domain and peaking at 538 nm, with a shoulder at $\lambda = 460$ nm (Figure 3). In the presence of DBU (up to 4 equiv.), the complete disappearance of **8** is monitored in favor of the formation of neutral quinoidal form **9**, exhibiting a broad blue-shifted absorption in the blue region at $\lambda = 460$ nm. Molecule **8** is also emissive (fluorescence) at $\lambda_{em} = 621$ nm in MeCN (excitation at $\lambda = 560$ nm) (Figure 3). Upon addition of DBU, the fluorescence is blue-shifted to a dual emission at 552 and 585 nm (formation of **9**) and significantly quenched (excitation at $\lambda = 475$ nm). These optical data are consistent with similar triamino-phenaziniums recently reported in the literature [41], except for the presence of a dual emission for **9**, due probably to excited-state intramolecular proton transfer [42].

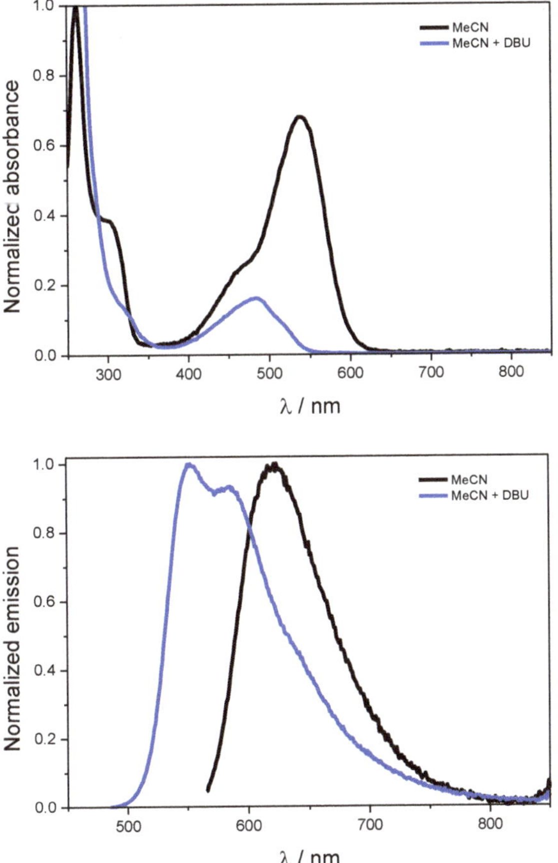

Figure 3. Normalized absorption (**top**) and emission (**bottom**) spectra of **8** (black) and **9** (blue, after addition of DBU) in MeCN.

3. Discussion

Molecule **8** belongs to the induline 3B family **10**, which has a long history in the field of textile and paint pigments [43–45]. Although described in 1923 [43,44], induline 3B (**10**) was little studied until 2012, when a study by Roy et al. revisited its chemistry to describe various N-aryl substituted analogues using a similar approach (self-condensation of aniline derivatives) [43].

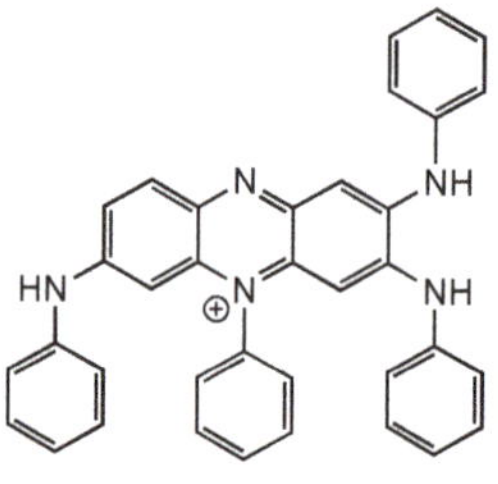

10

When we started to screen reaction conditions with the hope of finding an experimental procedure for the hydrogenation of azacalixarene **6**, we first avoided the use of acids and used hydrogenation on Pd/C and direct air-oxidation which led to molecule **7**. Previous works on macrocyclic chemistry reported that in the presence of acids, possible competition between degradation and hydrogenation might occur [46]. Here, we screened conditions for hydrogenation in acidic medium and we indeed observed a competing ring-opening reaction. The mechanism by which the macrocycle **5** breaks up is difficult to determine but it is wise to hypothesize a breakup within the corresponding azacalixquinarene **A** bearing amino/iminium bridges to form intermediate **B**. The protonation of this latter would afford **C**, which can then undergo a transamination reaction to form a tricyclic intermediate **D**, which will aromatize to **8** by proton migration (Scheme 2).

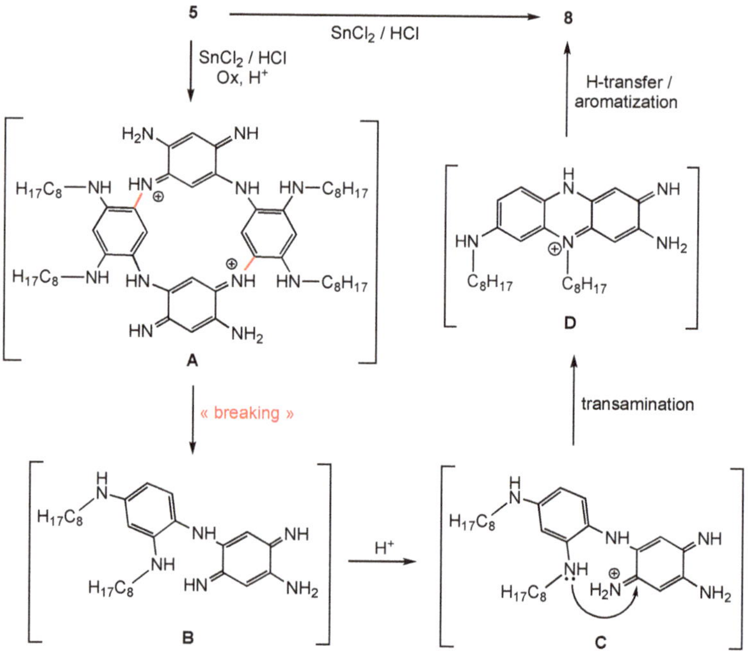

Scheme 2. Suggested mechanism to convert **5** into **8**.

To investigate the key role of the oxidizing agent and of the N-substituents, we considered the same reaction from the unsubstituted analogue **11** [38] in the absence of air. Remarkably, its reduction using $SnCl_2 \cdot 2H_2O$ (32 equiv.) in the presence of HCl (12N) (same conditions as for **5**) but in a sealed tube (preventing oxidation steps) furnished the macrocycle **12•nHCl** as a yellow powder (no ring opening, Scheme 3). The degree of protonation of **12** was impossible to determine but it is reasonable to consider the protonation of at least 4 sites in order to reduce the electron density of the aromatic rings and consequently prevent oxidation of the tetraaminobenzene units.

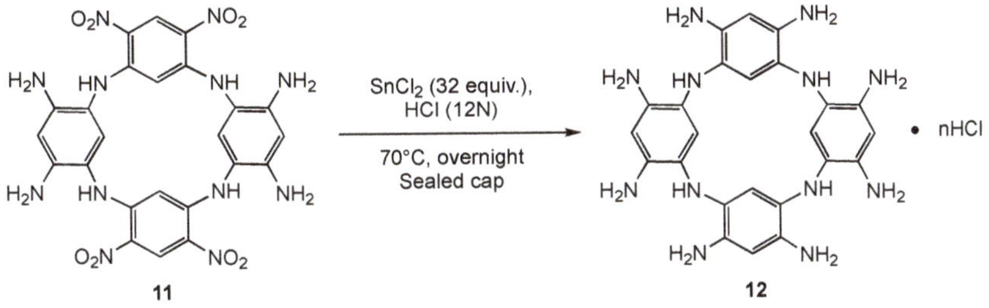

Scheme 3. Synthesis of **12•nHCl**.

This degree of protonation was confirmed by the ^{1}H NMR proton of **12•nHCl** in D_2O which shows only two signals at 6.31 and 7.28 ppm (see Figure S4, Supplementary Materials) in agreement with a highly symmetrical system in solution. Interestingly, when **12•nHCl** (as powder) was mixed with 10 mL of concentrated HCl and dropwise deionized water was added to dissolve all the solid under argon flux, spherical aggregates were formed after four days (see Figure S3).

In order to highlight the importance of the heteroatom (N vs. O) at the periphery in the reactivity of the azacalixarene, we next considered the reduction of compound **15** under the same conditions. Tetrahydroxy-tetranitro-azacalixarene **15** was first synthesized by condensation of the commercially available diaminoresorcinol **13** and difluorodinitro-benzene **14** (1 equiv.) in THF. After stirring during two days, the resulting solid in suspension was isolated by filtration affording macrocycle **15** in one step as a brown solid in a 70% yield (Scheme 4).

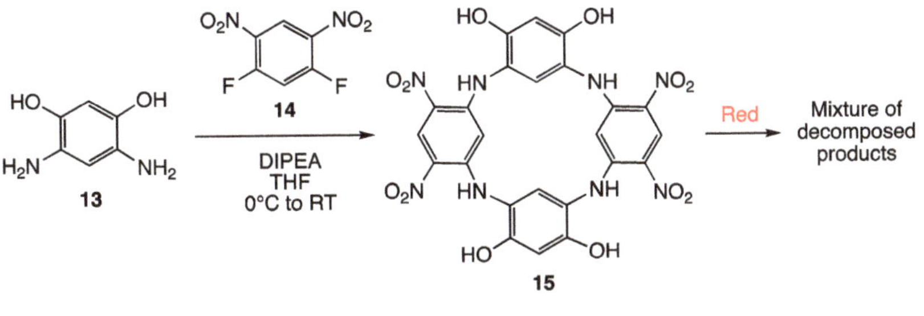

Scheme 4. Synthesis of tetrahydroxy-tetranitroazacalixarene **15** and its reduction by using procedures similar to those for **5**.

The ^{1}H NMR spectrum of **15** shows an unusual high-field chemical shift of the intra-annular aromatic protons at δ 5.70 ppm that suggests a 1,3-alternate conformation, in which these protons are located inside the anisotropic shielding. This observation confirms the high stability of this specific conformation in tetranitrocalixarenes even in the presence of H donor/acceptor sites inherent to the presence of the OH groups. The structure of **15** could be fully established by X-ray analysis, which confirmed its 1,3-alternate conformation stabilized by intramolecular hydrogen bonds involving the protons of the NH bridges of the macrocycle and the NO$_2$ groups, but also the participation of two hydroxy functions which interact with the nitrogen atom of the bridge (Figure 4). One of the most striking features of this structure is the sp^2 hybridization adopted by all the nitrogen bridging atoms that are conjugated with their adjacent dinitrobenzene rings, as already observed in related tetranitro-azacalixarenes [36].

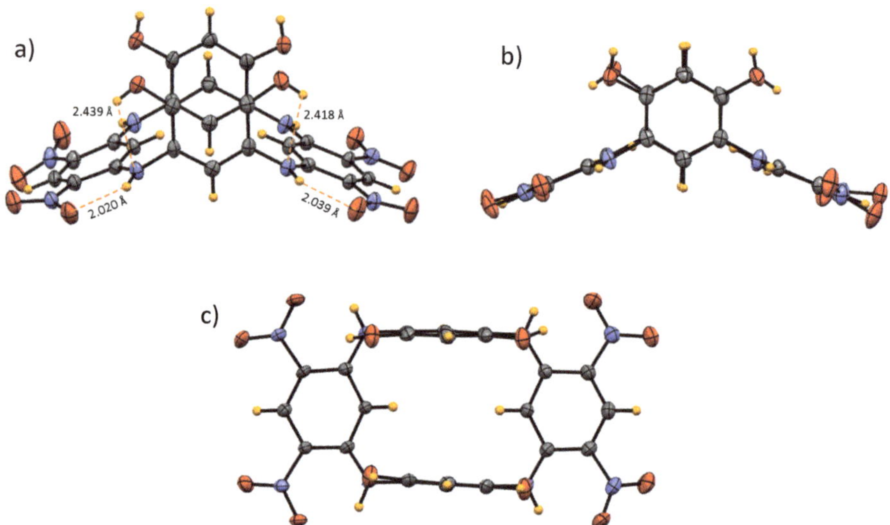

Figure 4. Three views of the single-crystal X-ray structure of **15**; (**a**) H-bonding interaction, (**b**) side view, (**c**) top view.

We then considered the reduction of compound **15** as for **5** as well as using many others using alternative reducing agents and/or conditions. In all cases, a full degradation of the reaction products, giving a complex mixture whose constituents could not be identified. This observation clearly confirms the key role of the amino functions at the periphery in the breaking of azacalixarenes into induline derivatives.

4. Materials and Methods

All reagents and solvents were purchased and used without purification. Liquid Nuclear Magnetic Resonance spectra were recorded on a Brucker advance 250 MHz (250 MHz for ^{1}H NMR and 62 MHz for ^{13}C) or on a JOEL Eclipse 400 MHz spectrometer. Chemical shifts are given in ppm relative to the signal (residual peak) of the solvent. The abbreviations used correspond to br = broad, s = singlet, d = doublet, t = triplet, and m = multiplet. Solid ^{13}C NMR were recorded on a Bruker Avance III WB 400 spectrometer. UV-Visible spectra were recorded at room temperature with a Varian Cary 50 UV-Vis spectrophotometer. High-resolution mass spectrometry (HRMS) analysis was performed by the "Spectropole" of Aix-Marseille University. Absorptions were performed in quartz cells of 1 cm and 1 mm. Emission spectra were measured using a Horiba-Jobin Yvon Fluorolog-3 spectrofluorometer equipped with a three-slit double-grating excitation and a spectrograph emission monochromator with dispersions of 2.1 nm mm-1 (1200 grooves per mm). A 450 W xenon continuous wave lamp provided excitation. The luminescence of diluted solutions was detected at right angle using 10 mm quartz cuvettes.

Compound **8**: to a solution of compound **5** (50 mg, 0.048 mmol, 1.0 eq) in absolute EtOH (50 mL), $SnCl_2 \cdot 2H_2O$ (343 mg, 1.52 mmol, 32 eq) and HCl (12N, 0.13 mL) were added. The mixture was allowed to reflux overnight, then neutralized with $NaHCO_3$ before the addition of EtOH (30 mL) and water (20 mL). After concentration, the residue was extracted with DCM/EtOH (3/1, *v/v*). The red organic phase was washed with an aqueous solution of HPF_6 (1%*w/w* in water, 4 × 150 mL) and brine (100 mL), dried with $MgSO_4$ and concentrated under vacuum to afford **8** as a dark red solid (35 mg, 0.059 mmol, 62% yield). ^{1}H NMR (400 MHz, CD_3OD): δ 7.68 (d, *J* = 9.2 Hz, 1H), 7.17 (d, *J* = 8.8 Hz, 1H),

6.99 (s, 1H), 6.93 (s, 1H), 6.45 (br s, 1H), 4.53 (t, J = 7.6 Hz, 2H), 3.29 (m, 2H), 1.91 (quint, J = 7.2 Hz, 2H), 1.74 (quint, J = 7.2 Hz, 2H), 1.63 (quint, J = 7.2 Hz, 2H), 1.47–1.23 (m, 18H), 0.93–0.88 (m, 6H). ^{13}C NMR (100 MHz, CD$_3$OD): 154.8, 152.1, 141.1, 139.6, 136.8, 134.2, 132.8, 132.1, 121.6, 108.8, 93.7, 90.3, 44.4, 33.1, 33.1, 30.7, 30.6, 29.8, 28.5, 28.0, 27.6, 23.8, 14.5. HRMS (ESI-TOF): m/z [M + NH$_4$]$^+$ for C$_{28}$H$_{44}$N$_5{}^+$ calcd. 450.3591, found 450.3592, err. < 1 ppm.

Compound **12•nHCl**: precursor **11** (50 mg) and SnCl$_2$ (500 mg, 32 eq) were mixed in a screw-cap vial with 10 mL of 37% HCl. The vial was then sealed with a Teflon-lined cap and placed in an oil bath at 70 °C with agitation for 20h. The clear yellow suspension was then cooled to room temperature. A volume of 40 mL of HCl (12N) was then mixed in the suspension and placed in an ultrasound bath for 10 min. The resulting solid was collected by filtration and washed successively with a mixture of MeCN and conc. HCl (40 mL/10 mL) and 2 × 20 mL of Et$_2$O. The crude vanilla powder **12•nHCl** (70 mg) was dried by flux of argon and stored at −20 °C. ^{1}H NMR (250 MHz, D$_2$O): 7.28 (s, 4H), 6.31 (s, 4H). Solid state ^{13}C NMR spectrum: 130–100 (m, sp^2 C-C), 132–148 (m, sp^2 C-N). Further characterization could not be carried out because of its high instability.

Compound **15**: to a solution of 1,5-difluoro-2,4-dinitrobenzene **14** (172.5 mg, 0.845 mmol, 0.9 eq) in THF (45 mL), 4,6-diaminoresorcinol dihydrochloride **13** (200 mg, 0.939 mmol, 1.0 eq) was added. The flask was sealed and degassed by 3 times of pump-argon cycling. Then, under an inert atmosphere, degassed *N*,*N*-diisopropylethylamine (DIPEA) (1.3 mL, 7.51 mmol, 8.0 eq) was added dropwise by a syringe at 0 °C. The solution was maintained at 0 °C for 6 h, and then allowed to warm at room temperature for 2 days. The resulting solid in suspension was isolated by filtration and washed with HCl (6N), EtOH and Et$_2$O to afford the desired product **15** as an orange solid (177.7 mg, 0.292 mmol, 70% yield). ^{1}H NMR (400 MHz, DMSO-d6): 9.71 (br s, 4H), 9.12 (br s, 4H), 9.01 (s, 2H), 6.83 (s, 2H), 6.51 (s, 2H), 5.50 (s, 2H). 13C NMR (100 MHz, DMSO-d6): 153.7, 148.9, 129.4, 128.0, 124.4, 115.8, 104.2, 94.0. HRMS (ESI-TOF): m/z [M + H]$^+$ for C$_{24}$H$_{17}$N$_8$O$_{12}{}^+$ calcd. 609.0960, found 609.0956, err. < 1 ppm.

5. Conclusions

We showed that depending on the reduction conditions, tetranitroazacalixarene **5** is at the crossroad of two different families of dyes including triamino-phenazinium **8** that results from a competing ring-opening reaction. The mechanism by which the macrocycle **5** breaks up is difficult to determine but we were able to highlight the key role of the amino functions at the periphery of the macrocycle and the need of an oxidizing agent (air). We are aware that the degradation of a sophisticated macrocycle (**5**) is not an efficient strategy to access induline derivatives by comparison with the straightforward synthesis recently described based on stepwise nucleophilic aromatic substitutions [41]. This work is more concerned with describing the versatility of azacalixarene-type macrocycles, beyond their "classical" use in supramolecular chemistry.

Supplementary Materials: The following supporting information can be downloaded at: https://www.mdpi.com/article/10.3390/molecules28248113/s1. ^{1}H and ^{13}H NMR data. Crystal data for complex **15** has been deposited in the CCDC database as deposition number 2308474 and can be accessed from there. The cif file for **15** is available as supplementary information on request.

Author Contributions: Conceptualization, Z.C. and O.S.; methodology, validation, formal analysis, Z.C., G.C., O.G., B.M. and O.S.; investigation, Z.C. and O.S.; writing—original draft preparation, writing—review and editing; supervision, project administration, funding acquisition, O.S. All authors have read and agreed to the published version of the manuscript.

Funding: This work was supported by the Centre National de la Recherche Scientifique, the Ministère de l'Enseignement supérieur et la Recherche (France).

Institutional Review Board Statement: Not applicable.

Informed Consent Statement: Not applicable.

Data Availability Statement: Data are contained within the article and Supplementary Materials.

Acknowledgments: O.S. wishes to thank Michel Giorgi (Spectropole, Marseille) for X-ray analysis.

Conflicts of Interest: The authors declare no conflict of interest.

References

1. Baldini, L.; Casnati, A.; Sansone, F.; Ungaro, R. Calixarene-based multivalent ligands. *Chem. Soc. Rev.* **2007**, *36*, 254–266. [CrossRef]
2. Kim, J.S.; Quang, D.T. Calixarene-Derived Fluorescent Probes. *Chem. Rev.* **2007**, *107*, 3780–3799. [CrossRef]
3. Ikeda, A.; Shinkai, S. Novel Cavity Design Using Calix[n]arene Skeletons: Toward Molecular Recognition and Metal Binding. *Chem. Rev.* **1997**, *97*, 1713–1734. [CrossRef]
4. Neri, P.; Sessler, J.L.; Wang, M.-X. *Calixarenes and Beyond*, 1st ed.; English Edition; Springer: Berlin/Heidelberg, Germany, 2016; 1653p.
5. Lhoták, P. Chemistry of Thiacalixarenes. *Eur. J. Org. Chem.* **2004**, *2004*, 1675–1692. [CrossRef]
6. Morohashi, N.; Narumi, F.; Iki, N.; Hattori, T.; Miyano, S. Thiacalixarenes. *Chem. Rev.* **2006**, *106*, 5291–5316. [CrossRef] [PubMed]
7. Kumar, R.; Lee, Y.O.; Bhalla, V.; Kumar, M.; Kim, J.S. Recent developments of thiacalixarene based molecular motifs. *Chem. Soc. Rev.* **2014**, *43*, 4824–4870. [CrossRef] [PubMed]
8. Botha, F.; Eigner, V.; Dvořákovác, H.; Lhoták, P. Arylation of thiacalix[4]arenes using organomercurial intermediates. *New J. Chem.* **2016**, *40*, 1104–1110. [CrossRef]
9. Mikšátko, J.; Eigner, V.; Lhoták, P. Unexpected cleavage of thiacalix[4]arene sulfoxides. *RSC Adv.* **2017**, *7*, 53407–53414. [CrossRef]
10. Maes, W.; Dehaen, W. Oxacalix[n](het)arenes. *Chem. Soc. Rev.* **2008**, *37*, 2393–2402. [CrossRef] [PubMed]
11. Cottet, K.; Marcos, P.M.; Cragg, P.J. Fifty years of oxacalix[3]arenes: A review. *Beilstein J. Org. Chem.* **2012**, *8*, 201–226. [CrossRef] [PubMed]
12. Upadhyay, H.; Harikrishnan, U.; Bhatt, D.; Dhadnekar, N.; Kumar, K.; Panchal, M.; Trivedi, P.; Sindhav, G.; Modi, K. A Highly Selective Pyrene Appended Oxacalixarene Receptor for MNA and 4-NP Detection: An Experimental and Computational Study. *J. Fluoresc.* **2023**. [CrossRef] [PubMed]
13. Zerr, P.; Mussrabi, M.; Vicens, J. Isolation and characterization of a new oxacalixarene. *Tetrahedron. Lett.* **1991**, *32*, 1879–1880. [CrossRef]
14. Smith, G.W. Crystal Structure of a Nitrogen Isostere of Pentacyclo-Octacosadodecaene. *Nature* **1963**, *198*, 879. [CrossRef]
15. Ito, A.; Ono, Y.; Tanaka, K. N-Methyl-Substituted Aza[1n]metacyclophane: Preparation, Structure, and Properties. *J. Org. Chem.* **1999**, *64*, 8236–8241. [CrossRef]
16. Tsue, H.; Ishibashi, K.; Tamura, R. Azacalixarene: A New Class in the Calixarene Family. *Topic Heterocycl. Chem.* **2008**, *17*, 73–96. [CrossRef]
17. Tsue, H.; Ishibashi, K.; Tokita, S.; Matsui, K.; Takahashi, H.; Tamura, R. Frozen 1,3-Alternate Conformation of Exhaustively Methylated Azacalix[4]arene in Solution: Successful Immobilization by Small but yet Sufficiently Bulky O-Methyl Group. *Chem. Lett.* **2007**, *36*, 1374–1375. [CrossRef]
18. Tsue, H.; Ishibashi, K.; Takahashi, H.; Tamura, R. Exhaustively Methylated Azacalix[4]arene: Preparation, Conformation, and Crystal Structure with Exclusively CH/π-Controlled Crystal Architecture. *Org. Lett.* **2005**, *7*, 2165–2168. [CrossRef]
19. Vale, M.; Pink, M.; Rajca, S.; Rajca, A. Synthesis, Structure, and Conformation of Aza[1n]metacyclophanes. *J. Org. Chem.* **2008**, *73*, 27–35. [CrossRef]
20. Xuea, M.; Chen, C.-F. Aromatic single-walled organic nanotubes self-assembled from NH-bridged azacalix[2]triptycene[2]pyridine. *Chem. Commun.* **2011**, *47*, 2318–2320. [CrossRef]
21. Tsue, H.; Ishibashi, K.; Tokita, S.; Takahashi, H.; Matsui, K.; Tamura, R. Azacalix[6]arene Hexamethyl Ether: Synthesis, Structure, and Selective Uptake of Carbon Dioxide in the Solid State. *Chem. Eur. J.* **2008**, *14*, 6125–6134. [CrossRef]
22. Yi, Y.; Fa, S.; Cao, W.; Zeng, L.; Wang, M.; Xu, H.; Zhang, X. Fabrication of well-defined crystalline azacalixarene nanosheets assisted by Se···N non-covalent interactions. *Chem. Commun.* **2012**, *48*, 7495–7497. [CrossRef] [PubMed]
23. Caio, J.M.; Esteves, T.; Carvalho, S.; Moiteiro, C.; Félix, V. Azacalix[2]arene[2]triazine-based receptors bearing carboxymethyl pendant arms on nitrogen bridges: Synthesis and evaluation of their coordination ability towards copper(ii). *Org. Biomol. Chem.* **2014**, *12*, 589–599. [CrossRef]
24. Ishibashi, K.; Tsue, H.; Takahashi, H.; Tamura, R. Azacalix[4]arene tetramethyl ether with inherent chirality generated by substitution on the nitrogen bridges. *Tetrahedron Asymmetry* **2009**, *20*, 375–380. [CrossRef]
25. Sakamaki, D.; Ito, A.; Matsumoto, T.; Tanaka, K. Electronic structure of tetraaza[1.1.1.1]o,p,o,p-cyclophane and its oxidized states. *RSC Adv.* **2014**, *4*, 39476–39483. [CrossRef]
26. Sakamaki, D.; Ito, A.; Furukawa, K.; Kato, T.; Tanaka, K. Meta–Para-Linked Octaaza[18]cyclophanes and Their Polycationic States. *J. Org. Chem.* **2013**, *78*, 2947–2956. [CrossRef] [PubMed]
27. Yan, X.Z.; Pawlas, J.; Goodson, T.; Hartwig, J.F. Polaron Delocalization in Ladder Macromolecular Systems. *J. Am. Chem. Soc.* **2005**, *127*, 9105–9116. [CrossRef] [PubMed]
28. Ito, A.; Ono, Y.; Tanaka, K. The Tetraaza[1.1.1.1]m,p,m,p-cyclophane Dication: A Triplet Diradical Having Two m-Phenylenediamine Radical Cations Linked by Twisted Benzenes. *Angew. Chem. Int. Ed.* **2000**, *39*, 1072–1075. [CrossRef]

29. Ishibashi, K.; Tsue, H.; Sakai, N.; Tokita, S.; Matsui, K.; Yamauchi, J.; Tamura, R. Azacalix[4]arene cation radicals: Spin-delocalised doublet- and triplet-ground states observed in the macrocyclic m-phenylene system connected with nitrogen atoms. *Chem. Commun.* **2008**, 2812–2814. [CrossRef]

30. Kulszewicz-Bajer, I.; Maurel, V.; Gambarelli, S.; Wielgus, I.; Djurado, D. Ferromagnetic spins interaction in tetraaza- and hexaazacyclophanes. *Phys. Chem. Chem. Phys.* **2009**, *11*, 1362–1368. [CrossRef] [PubMed]

31. Bujak, P.; Kulszewicz-Bajer, I.; Zagorska, M.; Maurel, V.; Wielgus, I.; Pron, A. Polymers for electronics and spintronics. *Chem. Soc. Rev.* **2013**, *42*, 8895–8999. [CrossRef]

32. Touil, M.; Lachkar, M.; Siri, O. Metal-free synthesis of azacalix[4]arenes. *Tetrahedron Lett.* **2008**, *49*, 7250–7252. [CrossRef]

33. Haddoub, R.; Touil, M.; Raimundo, J.-M.; Siri, O. Unprecedented Tunable Tetraazamacrocycles. *Org. Lett.* **2010**, *12*, 2722–2725. [CrossRef]

34. Canard, G.; Andeme Edzang, J.; Chen, Z.; Chessé, M.; Elhabiri, M.; Giorgi, M.; Siri, O. Alternate Tetraamido-Azacalix[4]arenes as Selective Anion Receptors. *Chem. Eur. J.* **2016**, *22*, 5756–5766. [CrossRef]

35. Tsue, H.; Oketani, R. Azacalixarene: An Ever-Growing Class in the Calixarene Family. In *Advances in Organic Crystal Chemistry*; Springer: Tokyo, Japan, 2015; pp. 241–261. ISBN 978-4-431-55554-4.

36. Konishi, H.; Hashimoto, S.; Sakakibara, T.; Matsubara, S.; Yasukawa, Y.; Morikawa, O.; Kobayashi, K. Synthesis and conformational properties of tetranitroazacalix[4]arenes. *Tetrahedron Lett.* **2009**, *50*, 620–623. [CrossRef]

37. Pascal, S.; Lavaud, L.; Azarias, C.; Varlot, A.; Canard, G.; Giorgi, M.; Jacquemin, D.; Siri, O. Azacalixquinarenes: From Canonical to (Poly-)Zwitterionic Macrocycles. *J. Org. Chem.* **2019**, *84*, 1387–1397. [CrossRef]

38. Chen, C.; Giorgi, M.; Jacquemin, D.; Elhabiri, M.; Siri, O. Azacalixphyrin: The Hidden Porphyrin Cousin Brought to Light. *Angew. Chem. Int. Ed.* **2013**, *52*, 6250–6254. [CrossRef]

39. Chen, Z.; Haddoub, R.; Mahé, J.; Marchand, G.; Jacquemin, D.; Edzang, J.A.; Canard, G.; Ferry, D.; Grauby, O.; Ranguis, A.; et al. N-Substituted Azacalixphyrins: Synthesis, Properties, and Self-Assembly. *Chem. Eur. J.* **2016**, *22*, 17820–17832. [CrossRef]

40. Seillan, C.; Marsal, P.; Siri, O. New class of highly stable nonaromatic tautomers. *Org. Biomol. Chem.* **2010**, *8*, 3882–3887. [CrossRef]

41. Chen, Z.; Pascal, S.; Daurat, M.; Lichon, L.; Nguyen, C.; Godefroy, A.; Durand, D.; Ali, L.M.A.; Bettache, N.; Gary-Bobo, M.; et al. Modified Indulines: From Dyestuffs to in vivo Theranostic Agents. *ACS Appl. Mater. Interfaces* **2021**, *13*, 30337–30349. [CrossRef] [PubMed]

42. Yushchenko, D.A.; Shvadchak, V.V.; Klymchenko, A.S.; Duportail, G.; Mély, Y.; Pivovarenko, V.G. 2-Aryl-3-hydroxyquinolones, a new class of dyes with solvent dependent dual emission due to excited state intramolecular proton transfer. *New J. Chem.* **2006**, *30*, 774–781. [CrossRef]

43. Roy, S.K.; Samanta, S.; Sinan, M.; Ghosh, P.; Goswami, S. Aerial Oxidation of Protonated Aromatic Amines. Isolation, X-ray Structure, and Redox and Spectral Characteristics of N- Containing Dyes. *J. Org. Chem.* **2012**, *77*, 10249–10259. [CrossRef] [PubMed]

44. Berneth, H. *ULLMANN'S Encyclopedia of Industrial Chemistry*; Wiley-VCH Verlag GmbH & Co. KGaA: Weinheim, Germany, 2012.

45. Solodar, W.E.; Monahan, A.R. The synthesis and Spectroscopic Characterization of Indulin 6B Tetrasulfonate. *Can. J. Chem.* **1976**, *54*, 2909–2914. [CrossRef]

46. Journot, G.; Neier, R.; Gualandi, A. Hydrogenation of Calix[4]pyrrole: From the Formation to the Synthesis of Calix[4]pyrrolidine. *Eur. J. Org. Chem.* **2021**, *2021*, 4444–4464. [CrossRef]

MDPI
St. Alban-Anlage 66
4052 Basel
Switzerland
www.mdpi.com

Molecules Editorial Office
E-mail: molecules@mdpi.com
www.mdpi.com/journal/molecules